DATE DUE

DEC 20 1980		
MAR 8 1981		
23 1991		
OCT 28 1991		
MAY 1992		
GAYLORD		PRINTED IN U.S.A.

ALISM

PRENTICE-HALL METHODS AND THEORIES
IN THE SOCIAL SCIENCES SERIES

EDITORS

Herbert L. Costner
Neil J. Smelser

FUNCTIONALISM

Mark Abrahamson

University of Connecticut

PRENTICE-HALL, INC., Englewood Cliffs, New Jersey 07632

Library of Congress Cataloging in Publication Data

ABRAHAMSON, MARK.
  Functionalism.

  Includes bibliographical references and index.
  1. Functional analysis (Social sciences) I. Title.
GN363.A27   301'.01'5157   77-6828
ISBN 0-13-331900-8

Printed in the United States of America

10 9 8 7 6 5 4 3 2 1

Prentice-Hall International, Inc., *London*
Prentice-Hall of Australia Pty. Limited, *Sydney*
Prentice-Hall of Canada, Ltd., *Toronto*
Prentice-Hall of India Private Limited, *New Delhi*
Prentice-Hall of Japan, Inc., *Tokyo*
Prentice-Hall of Southeast Asia Pte. Ltd., *Singapore*
Whitehall Books Limited, *Wellington, New Zealand*

contents

6

Assessing Functional Interpretations   93

Index   111

preface

The content of most of my theory courses while a student in the late 1950s and early 1960s, was functionalist in perspective. I did not realize this at the time, however; I merely regarded what I was learning as theory. The term functionalism meant little to me. As I recall, I equated the term with Parsons and found most of the debate about Parsons to be boring. (C. Wright Mills's criticisms accurately expressed my sentiments.)

The turmoil of the late 1960s was, of course, reflected in sociology and the "establishment" in the discipline was attacked, along with establishments everywhere else. For me, it was initially a revelation to discover that functionalism was regarded as the "theoretical establishment." Reading the critiques of functionalism was enormously helpful, because it helped me grasp some similarities among a number of theorists whose writings had influenced me. It also created some personal dissonance because the professor who had the greatest impact on my tastes in theory—in both undergraduate and graduate courses—was Alvin Gouldner; and he was now among the major critics of functionalism.

As I wrestled with the debates in the discipline, I became very soured by the attacks on functionalism, which I was hearing from younger colleagues and from graduate students. They mouthed the popularized deficiencies of functionalism, but with seemingly little understanding of what they were criticizing. My immediate response was to offer a graduate seminar in functionalism. The course was well attended and well received, more enthusiastically than I had expected. I truly enjoyed observing students as they read and discussed the classical functionalist writings, and their serious contemplation of the criticisms. I felt that a balanced understanding emerged—one that involved both a critical and appreciative stance toward functionalism. Over the next few years I wrote this book, trying to build that balance into it while maintaining a level that could be read and understood by undergraduate students as well.

I am indebted to a number of people who helped along the way: Ephraim Mizruchi, Manfred Stanley, Neil Smelser, and Herbert Costner read portions of the manuscript and offered insightful suggestions—I did not always listen to their advice, however, so I alone am responsible for the content; my editor at Prentice-Hall, Edward Stanford, was helpful and encouraging; and I am also appreciative of the secretarial help I received at Syracuse, Georgia State, and Connecticut, the universities at which I wrote this book. But most of all, thanks to Mar, who was always there.

Mark Abrahamson
Storrs, Connecticut

FUNCTIONALISM

I
The Functionalist Perspective

1

systems and functions

Have you ever heard a child ask a parent, "Why do people have stomachs?" Reflect on the question for a moment. How would you answer it? Chances are you would answer by pointing out what would happen if someone did not have a stomach. Thus, you might respond to the question with another question: "Where would the food go if we didn't have stomachs?" If you were to elaborate a bit, you might explain the idea of the digestive system, and you might also describe the specific components involved. Then you could concentrate on the role of the stomach in the digestive system, and the role of the digestive system in sustaining the life of the organism.[1]

I cite this answer of why people have stomachs because it is essentially a functional explanation. The basic characteristic of such explanations is the utilization of a system model and an emphasis upon the contributions of the elements to the maintenance of the system. At an abstract level this pattern

---

[1]The question "why?" is often a request to specify necessary causes; that is, why something must be, or happen, the way that it does. The type of explanation offered above for the stomach, which is functional, does not specify cause in the usual sense of the term. See Ernest Nagel, *The Structure of Science* (New York: Harcourt Brace, 1961). The distinction between function and cause will be examined in later chapters.

may be difficult to follow, so let us take a closer look at this characteristic of functionalism. The digestive system, to be more specific, contains elements such as: saliva, alimentary canal, stomach, and intestines. The system is comprised of both these components and the relationships among them. Thus, a system is more than the sum of its parts; it is also the relationships among its parts.

It is important to realize that specific systems are often nothing more than mental constructs, inferred for heuristic purposes; that is, by viewing phenomena as elements of a system, one creates what will hopefully provide a useful theoretical paradigm with which to analyze the phenomena. In other words, there may be no compelling reason in reality to view the phenomena as part of a specific system, but doing so may produce certain insights, as it did in clarifying the role of the stomach. On the other hand, it may put phenomena together in arbitrary ways that obscure their analysis. For example, the absorption of oxidized food into the body is often presented as the task of the digestive system. However, the complete process involves not only those elements and processes conventionally thought to constitute the digestive system, but the elements and processes that guide the circulation of compounds throughout the body via the blood stream as well. By conventional definitions, though, the blood stream is an element of the circulatory system. Arbitrary system boundaries may therefore be obscuring some aspects of processes while, at the same time, clarifying others.

Every abstract conception has this quality of illuminating while obscuring. The notion of systems, for example, has provided a concept that has helped to clarify processes in many different areas of analysis. Thus, a system model in economics sensitizes researchers to expect a relationship between wholesale prices and inflation; it alerts sociologists to expect a relationship between work roles and family interaction; and so on. However, there are two problems that may develop due to these insights. First, alternative conceptions, which might be better, tend to be disregarded, and that can be a very costly move. This is an obvious cost that requires no elaboration. The second type of cost involves the proverbial "wild goose chase" and it is a more subtle, but very recurrent, cost. For example, a system model has led biological researchers to continually study the role of the human appendix. They reasoned that it must contribute to the digestive system, and they proceeded to study with this idea in mind; but their search was in vain. Periodically, non-digestive roles are also tentatively attributed to the appendix; for example, this organ increases a person's resistance to the common cold. Despite the consistent finding of little supportive data, studies of the appendix continue. Perhaps this organ should simply be ignored; but the pervasive conception of systems continues to encourage speculation and research, which may be fruitless in the final analysis. Thus, a system model—like any other conception—involves a gamble: that the conception will illuminate more than it obscures.

It is important to study the models and conceptions that are pervasive in any field of inquiry, such as sociology, in order to understand the field itself. These notions not only shape the questions that are raised, they also suggest answers to the questions themselves. System notions have been very pervasive in sociology, and for the most part, such social systems have been viewed as functionally integrated. Hence, this book is devoted to an analysis of the implications and meanings of this conception.

Functional Systems

When a system has been conceptualized in functional terms, a number of assumptions are typically made concerning how such systems, and their component elements, operate. Among the more important assumptions are the following:

1) *The elements of a system are functionally interrelated.* This means that the normal operation of one element requires the normal operation of other elements. Within certain limits, however, one element is often viewed as being able to compensate for another. If food is not properly chewed, for example, it may disrupt the transmission capacities of the alimentary canal; but the canal may be able to accommodate for particles whose mastication is very incomplete. If two elements are able to overlap almost completely in their contributions to a system, they are termed "functional equivalents."

2) *The components of a system generally contribute positively to the continued operation of that system.* It is the analysis of such contributions, or consequences, that are of primary concern. It is this emphasis upon contributions, which are defined as functions, that earned functionalism its name. However, many theories also consider the possibility of elements that do not contribute positively to the perpetuation of the system of which they are a part. If the contribution is negative, that is, disruptive to the system, the element is termed "dysfunctional." There is also the possibility that an element makes no contribution at all. Such absence of effect is termed a "survival," implying that the element once made a positive contribution, but as the system changed, the contribution ceased without the element disappearing. Functional theories differ in their willingness to accept the possibility of survivals. Even where they are admitted, though, survivals tend to be seen only as temporary phenomena; that is, the functionless element has not disappeared—yet.

3) *Most systems affect other systems, and may be viewed as subsystems of the entire organism.* For example, the digestive system affects the circulatory system, and either may, in turn, be viewed as a subsystem, with the entire biological organism conceptualized as *the* system. The assumptions we have been noting are then made, whether the component being analyzed is an

element of a specific system, or is a subsystem viewed in relation to a larger system.

In addition, functional theories have tended to assume, at a more abstract level, that systems are highly integrated and stable. Associated with the assumption of integration are the further assumptions that the parts fit together "harmoniously," with little friction, and that systems are characterized by equilibrium; that is, basic relationships among components change little over time. More generally, fundamental changes of all types are viewed as sporadic at best, and as inherently resisted by systems. When changes do occur, given the integrated nature of the system's components, they are expected to have extensive ramifications. Specific functional theories have differed somewhat concerning these later points, however, and these variations introduce complexities that are best put off until Chapter 2.

Thus far, most of our examples have been drawn from biology, but it is apparent that similar types of thinking are prevalent in many disciplines. A solar system, for example, involves relationships among the sun and the planets that revolve around it. Similarly, an educational system is comprised of: age-graded schools, teaching staffs, parent associations, and the relationships among each. In all of these non-biological examples, a functionalist perspective again involves the same emphasis upon contributions and consequences within systems and between subsystems.

Functionalism, as we have seen, requires the prior conceptualization of a system before its "explanatory imagery" makes any sense. However, all systems do not need to be viewed from a functionalist perspective. Rather than emphasize the contributions of elements, for example, entire systems can be analyzed according to the kinds of continuous exchanges that characterize their relationships. Emphasis can be placed, therefore, upon transactions rather than functions. Correspondingly, systems may be analyzed without the assumption that they possess any of the typical functional attributes, such as integration or tendency toward equilibrium. In other words, system theories differ from each other according to the ways in which systems are seen to be organized, and a functional organization is only one of the possible alternatives.[2]

This book is, of course, about functionalism; and more specifically, about functionalism in sociology. It is an important theoretical approach to understand because, as we have seen, it has a very pervasive influence upon our thinking. Almost reflexively we organize our thoughts and search for answers within an implicitly functional systems mode. Now, every theoretical approach—including the conception of a system, however organized—has its costs in terms of alternatives foregone. That is, regardless of its usefulness, every approach has certain blind spots that might be illuminated by a different

[2]For further discussion of non-functional systems, see Walter Buckley, *Sociology and Modern Systems Theory* (Englewood Cliffs, N.J.: Prentice-Hall, 1967).

perspective. What, we will ask in this volume, is clarified and what is obscured by functionalism? For example, does the view of society as an "integrated organism" lead to a virtual disregard of social conflict?

Throughout most of the twentieth century, functional models have been dominant in sociological theory and research. Numerous theoretical approaches have recently been presented, however, with a polemical stance toward functionalism at their core; that is, their essential thrust is defined in juxtaposition to functionalism. Thus, the major theoretical arguments, counterattacks, and rejoinders in the field simply cannot be grasped without a prior understanding of functionalism.

The Development of System Perspectives

While systems need not be analyzed functionally, functionalism has been seen to require the prior conceptualization of a system. Therefore, to appreciate the history of functionalism, it is necessary to begin with the history of system theories. Sociology, as a distinct discipline, emerged well after system models had been developed in the physical and biological sciences. Initially, the fledgling discipline's perspectives were almost entirely shaped by the prevalent physical and biological systems models. This occured, in part, because disciplinary lines were not clearly drawn by contemporary standards, and the social theorists were often the natural science theorists as well. In any case, the history of functionalism in sociology most clearly begins with the rejection of Aristotelian "science" in the seventeenth century.

Aristotle's physics, for example, involved a conception of the universe with the earth at its center. The earth was also considered to have no natural, or inherent, motion. Unlike other planets and stars, it was "at rest." Motion, for Aristotle, was determined by the composition of any object, not by external forces or by the attractions of other objects. Objects that were heavy, such as rocks, would naturally fall down, he reasoned. Light weight objects, such as smoke, would naturally ascend up. Circular motion was seen as a characteristic of planets only, so that was attributed to their non-earthly composition, a composition whose natural movement was circular. (In addition, because circular movement has no point of origin or of termination, it was also congruent with Aristotle's assumption of perpetual celestial motion around the earth.)[3]

Note that Aristotle's simplified and incorrect view of the physical universe disregarded how the motion of an object related to the motion of other objects. As a theoretical model it differs sharply from contemporary concep-

---

[3]Bernard Cohen, *The Birth of a New Physics* (New York: Doubleday, 1960).

tions of a dynamic solar system in which components are seen as being in continuous interaction with each other. The emphasis upon motion as due to the inherent composition of an object is analogous to viewing an individual's behavior as the simple result of a genetic instinct rather than as a result of a combination of forces interacting within a personality system.

Copernicus's modified theories of matter and motion (1543), followed by the invention of the telescope (1609), were the precursors of Galileo's "new physics." His telescopic observations permitted him to study reflections of sunlight from the earth onto the moon. These observations led Galileo to argue that the earth was also a "wandering body," involved in a complex interaction with the rest of the universe. Because of the theological implications of this view of the universe, Galileo's theory did not put him in good standing in Padua's official circles, but his precise and detailed observations continued, despite this opposition.

Many of the specific laws Galileo presented were later found to be erroneous. However, the significance of Galileo's contribution to modern physics has to be seen as a breakthrough, in theoretical terms at least. It was the beginning of a conceptual reformulation that set the stage for Newton's still more impressive contributions in the latter part of the seventeenth century.[4] Basically, Galileo presented precise mathematical formulations of specific relationships. Newton built upon these formulations, and was able to recast physical thought into a dynamic view of the solar system, with multiple forces.

Also, during the early part of the seventeenth century in Padua, biological researchers were beginning to reject other aspects of Aristotelian thought, and they developed new models that were influenced by the development of mechanically-oriented system models. Technological discoveries were again crucial: in this case, the microscope (1624). However, while the seventeenth century revolution in physics primarily involved a new way of looking at old facts, the revolution in biology entailed more of an explosion of new facts.[5]

Aristotle had emphasized the importance of the heart in the human body and viewed it as analogous to the primacy of the sun in the cosmos: he saw both as the source of light-giving heat in their respective systems. With more detailed information at hand, the seventeenth century Paduan biologists likened the heart to a pump in a waterworks system, and recognized a more complex interaction between the heart and the circulation of blood. Thus, biological thought, like physical thought, was moving toward the conception of dynamic systems, though neither fully arrived at that point during the seventeenth century.

The influence of this mechanical systems approach did not stop with the

[4]Ernan McMullin, "Introduction," in *Galileo, Man of Science,* ed. Ernan McMullin (New York: Basic Books, 1967).

[5]Richard Westfall, *The Construction of Modern Science* (New York: Wiley, 1971).

biological sciences, however. It also provided one of the precursors to modern sociology, the social physics of the seventeenth century. In this view, society became "a new astronomical system whose elements were human beings."[6] Society, in this metaphor, was assumed to be subject analogously to the same laws and forces as other systems. Equilibrium, or balance, for example, was assumed to be characteristic of society as a result of inertia and gravitation (i.e., social attraction); political institutions were viewed as being built upon atoms (i.e., individuals) and molecules (i.e., groups); etc.

In the development of modern sociological theory it is also possible to differentiate between mechanistic and biological influences. In the mechanistic model, society is regarded as a more rationally constructed system whose components may be subjected to planned change. This type of approach is illustrated by the writing of Henri Saint-Simon, an eighteenth century social physicist, and founder of modern socialism. In his view, a new social order was emerging, and it was one in which science would be paramount. Political and administrative elites, for example, would rule on the basis of their expertise, and the irrational self-concepts of the feudal period would dissipate. As a result of these alterations, the new social system would be amenable to deliberate and planned changes.[7]

By contrast, the Scottish Moralists, eighteenth century social philosophers, were more influenced by the biological model. (Adam Smith and David Hume are probably the best known representatives of this school.) Social organization, according to the Scots, is the anomalous result of people acting in accordance with limited (i.e., selfish, hedonistic, or narrow) desires. They illustrated this phenomenon by analogy to the social life of bees. These insects coordinate their activity by flying through a series of geometric patterns. However, coordination is not their intent, and none of the bees understand the rules of geometry. They argued that people also follow narrow instincts or desires out of which an organized pattern emerges, such as is the case in commerce, law, etc. Rationality in human activity is correspondingly de-emphasized in this view, with the institutional components of social systems regarded as natural and spontaneous.[8]

Functionalism in modern sociology has been more influenced by the natural system models of biology than by the mechanistic and rational models. Thus, functional theories conventionally view system needs as inherent rather than intended; and deliberate, or enacted, change is seen as problematic, both in terms of frequency of occurence and probability of success. Despite these

[6]Pitrim A. Sorokin, *Contemporary Sociological Theories* (New York: Harper, 1928), p. 7.

[7]See the discussion of "rational" and "natural" system models in Alvin W. Gouldner, "Organizational Analysis," in *Sociology Today*, eds. Robert K. Merton, et al. (New York: Basic Books, 1959).

[8]See Louis Schneider, *The Scottish Moralists* (Chicago: The University of Chicago Press, 1967).

differences, however, the mechanical and biological models had congruent consequences in so far as both promoted a systems view of society. In addition, as the preceding discussion indicated, there was a clear historical interdependence between the biological and mechanical conceptions.

One final aspect of Aristotelian thought that was attacked in the seventeenth and eighteenth centuries was "vitalism," and its replacement by a growing naturalistic conception of events was particularly important to the development of a scientific study of society. Vitalism entails the belief that all living things possess a degree of autonomy; an autonomy that precludes their total determination by laws, and hence precludes their complete explanation or prediction. As defined by Catton, Aristotelian vitalism involves the belief that there is, "an inner sanctum whose door must remain forever barred to systematic research and explanation."[9] However, the impressive discoveries of biological and physical patterns, facilitated by the microscope and telescope, led to increasing confidence that all living things (including people and society) could be largely, if not totally, explained.

In summary, throughout the seventeenth century Aristotelian thought was attacked from all sides. If the attack is viewed as a scientific revolution, then the fruits of the revolution can be seen in the growing acceptance of complex and natural systems models, which were increasingly viewed as functionally organized. This aspect of the history of science has been succinctly summarized as follows:

> The modern reorientation of thought, the new models, appear to be centered in the *concept of system*. . . . Now we are presented, in all sciences, with . . . wholes, organizations, mutual interactions of many elements and processes, systems. . . .[10]

Classifying Observations

In utilizing the newly discovered instruments and techniques, scientists of the seventeenth century were soon overwhelmed by the burgeoning number of observations. Theoretical cogency lagged far behind the newly found empirical efficacy. Systems of classification, a first step in the developement of theory, were urgently needed, simply to reduce the data into more manageable categories.

[9]William B. Catton, Jr., *From Animistic to Naturalistic Sociology* (New York: McGraw-Hill, 1966), p. 30.

[10]Ludwig von Bertalanffy, "The Model of Open Systems," in *Biology, History and Natural Philosophy*, eds. Allen D. Breck and Wolfgang Yourgau (New York: Plenum Press, 1972).

By definition, a good system of classification will reduce the number of variables that have to be considered. It will also enhance the ability to find relationships among the types that are classified, but this second criterion was not emphasized until a century later by John Stuart Mill. Even in terms of data reduction, however, the seventeenth century classifications were not very efficient. A then-popular taxonomy of herbs, for example, described 6,000 different species. Most classifications of this period were inefficient, because they arbitrarily selected a single characteristic of plant or animal to generate a classification. There was little concern with the theoretical importance or usefulness of the single characteristic selected.[11]

In order to grasp the seriousness of the classification problem, let us consider a concrete issue, such as personality traits. A vast number of these traits can be identified: introversion, honesty, bravery, etc. In all, the English language contains about 18,000 names for such traits.[12] By constructing a classification of personality types, large numbers of specific traits can be subsumed under a much more limited number of categories. In a category such as simple schizophrenia, for example, it may be possible to include traits like apathy, withdrawal, and unresponsiveness.

Of course, all typologies are not equally useful. They differ in the number of total possible traits they can consider and, therefore, in their capacity for reducing the number of variables. They also differ in the ease of comparison that they permit among categorized types. Given 6,000 herbs or 18,000 personality traits, however, almost any classification must be of initial value in reducing the number of specific characteristics that have to be taken into account. In the long-run, a classification scheme based upon insignificant theoretical considerations may impede further development of theory; but until voluminous reams of observations are somehow reduced, theory development is highly improbable.

Throughout this century there has been a continuing tension in sociology, between schools that attempt to generate theories without regard to empirical data and schools that are devoted to the sheer generation of facts that are divorced from theory. If we accept as ideal a situation in which there is an interplay between theory and data—each influencing and also building upon the other—then it is fair to conclude that none of the schools described above, with their extreme emphasis upon either theory or data, have been successful.

Many of the most elaborate functional theories in sociology have had a regrettable lack of any data base; some of them will be described in the following chapter. The sheer generation of the facts approach in sociology was associated with the growing sophistication of survey research techniques during the 1930s. To a degree, survey research provided the twentieth century

---

[11]Westfall, *The Construction of Modern Science*. See Chapter 5.

[12]Gordon W. Allport and H.S. Odbert, "Trait Names," *Psychological Monographs*, 47 (1936).

sociological counterpart to the seventeenth century telescope and microscope. As sometimes utilized, it resulted in detailed, trivial observations about church attendance, voting preferences, and so on. Criticism of this approach was extensive. Its lack of substance was criticized by Mills, who termed it "abstracted empiricism." Merton labeled the approach, "radical empiricism," and stated that an advocate's motto might be: "this is demonstrably so, but we cannot indicate its significance."[13]

Classifications of data, for use in theory construction, have often been facilitated by functional models. Rather than focus upon isolated or trivial characteristics of elements, such classifications focus upon the contributions of elements to the systems of which they are a part. Thus, in the human body, numerous glands can be classified as endocrines based upon their common regulating function via internal secretions. Cutting across diverse disciplines, structural models have also been widely used as methods of developing classifications. And in many instances, structural and functional models have been used together.

Structure and Function

A simple approach to understanding the conception of structure is to view it as the organization of a phenonemon "in space." The valves and arteries of the heart, for example, have been structurally categorized according to their location, and divided into quadrants (e.g., right auricle, left ventricle). Similarly, spatial location can be utilized to categorize elements of a human community (e.g., suburbs, central business district). In each case, structure corresponds roughly with a photograph; it emphasizes the shape and form of a phenomenon at one point in time.

The equivalence between structure and a photograph, however, is an accurate portrayal only of some aspects of structural theories. Many of the best known structural theories sharply limit the role of empirical observation, and hence, limit the potential value of an actual photograph. "Real" or complete knowledge, it is contended, does not come from direct sensory contact because only manifestations of the real structure can be directly experienced. Real knowledge, by contrast, comes from mentally constructing concepts that permit the concrete phenomenon to be seen as manifestations of underlying structures. For example, Marx's analysis of the political economy emphasized its basically hidden (true) structure, and the way in which life was subtly determined by that structure. In an analysis similar in form to Marx's,

[13]Robert K. Merton, *Social Theory and Social Structure* (New York: Free Press, 1947), p. 83. See also, C. Wright Mills, *The Sociological Imagination* (New York: Oxford University Press, 1959).

Freud's examination of psychopathology emphasized the true, but hidden, structure of the unconscious mind. Structural theories, in each case, emphasize comprehension rather than precision; a glimpse of the real structure rather than exact detail about inconsequential phenomena.[14]

Functionalism has often been included, creating a structural-functional perspective, as theories have attempted to account for structural changes. At an empirical level, for example, the shape and form of the heart's valves are different in diastole than in systole; and activities also vary in different parts of the human community in the course of a day. Despite these changes, though, there is a pattern in which the same *forms* continuously reappear. The heart's form continuously changes from mode to mode; the form of the human community changes during the diurnal mode.

Structural-functionalism is created by assuming that if structure is organization in space, then function is organization in time. Any system, it is assumed, "must fuse the two . . . so that its function can maintain its structure."[15] Or, is it the other way around; namely, that its structure must maintain its function? Raising this question makes it apparent that structure and function present some aspects of a chicken and egg dilemma. Consider, for example, the two types of organization, space and time, in relation to the two principal sense modalities, sight and sound. The eye, Moulton has proposed, takes in a static view of the world in space. The ear, by contrast, provides information about a world that exists in time. Therefore, in listening to a symphony, its temporal organization—the melody—is best heard; its spatial organization—the score—is best seen.[16]

In reality, however, sight and sound can also be used as equivalents. It is possible to transpose between sense modalities; for example, hearing a symphony and visualizing its score. Thus, the distinction between spatial and temporal organization has its fuzzy areas, and the corresponding distinction between structure and function can also entail moot points. For the structural-functionalists who attempt to integrate the two, it is a manageable ambiguity. It must be pointed out, however, that all structural theories need not be functional in emphasis. The functional implications of underlying structures may be considered of equal importance, or may be largely dismissed as obvious or unimportant.[17] Correspondingly, the structures associated with functions may or may not be given equal attention by functionalists. In most instances there is some concern with both, but to highly varying degrees.

[14]See the introduction by Richard and Fernande De George, eds., *The Structuralists: From Marx to Levi-Strauss* (New York: Doubleday, 1972).

[15]F. Eugene Yates, et al., "Integration of the Whole Organism," in *Challenging Biological Problems,* ed. John A. Behnke (New York: Oxford University Press, 1972), p. 123.

[16]Adrian C. Moulton, *Structure, Function and Purpose* (New York: Liberal Arts, 1957). Moulton also makes an important distinction between objective and subjective time as a basis for organization. Note especially Chapter 5.

[17]See the discussion of ethnology and linguistics in Gunther S. Stent, "Limits to the Scientific Understanding of Man," *Science*, 187 (March 21, 1975).

In summary, seventeenth century natural science entailed a factual revolution, spurred by new empirical techniques. One ensuing theoretical development entailed the conception of systems, and most of the conceptualized systems came to be seen as functionally organized. The raw data of the seventeenth century also led to classification schemes, which were later based, in many instances, upon a conception of structure. These two historical processes are associated with two somewhat different types of functionalism in contemporary sociology: functionalism and structural-functionalism. The distinction between the two types is primarily a matter of degree, however, as most functional theories involve at least some concern with structure.

Male Initiation Ceremonies[18]

In order to clarify all the topics discussed to this point, it may be helpful to analyze all of them in relation to one specific problem, such as male initiation ceremonies. The following discussion is influenced by previously published studies,[19] but in order to illustrate the issues that are of particular concern to us, questions will be raised (and alternatives considered) that were not dealt with in the prior investigations.

Let us begin with the observation that in about one-third of all pre-industrial societies for which there is relevant data, "something special" happens to male youngsters between the ages of about ten and sixteen years. In most cases, when the "something special" does occur, it involves the boys being herded together in a compound and isolated from all females for several days. Tests of their endurance, hazing by adult males, and genital operations (e.g., circumcision) are also common. A substantial amount of ceremony and ritual typically occurs also, and it partly accounts for the varying ages of the boys (i.e., ten to sixteen years). Many of the ceremonies are expensive, involving events such as feasts and the entertaining of relatives. Because the costs are often the responsibility of the parents of the boys involved, villages often skip the ceremonies during some years, waiting until there is a sufficient number of boys within the customary age bracket. This lessens the costs for the individual parents.

Any theoretical analysis of these rituals initially faces a *classification*

[18]In the following pages we will examine only the initiation of males. Studies of female initiation practices suggest that they are related to substantially different conditions. See J. K. Brown, "A Cross Cultural Study of Female Initiation Ceremonies," *American Anthropologist,* 65 (1963).

[19]The most relevant study is John M. Whiting, Richard Kluckhohn, and Albert Anthony, "The Function of Male Initiation Ceremonies at Puberty," in *Readings in Social Psychology,* eds. Eleanor Macoby, et al. (New York: Holt, 1958). See also, Y. A. Cohen, "The Establishment of Identity in a Social Nexus," *American Anthropologist,* 66 (1964). Additional classifications of societies are presented in Robert B. Textor, *A Cross-Cultural Survey* (New Haven: HRAF Press, 1967).

problem: that is the need to conceptualize initiation ceremonies as a general type of event, such that specific societies can be similarly classified. Otherwise, given the diverse practices and diverse ages of participating youngsters, every society will appear to be unique. This is the previously discussed data reduction advantage of any classification. However, all classifications will not serve the data reduction task equally well. A relatively non-useful, but easy to develop approach would classify societies according to their types of hazing, mutilation, seclusion, and deprivation, along with other specific practices. It could result in an abundance of categories equal to the number of societies.

In order to reduce cases substantially, it is necessary to differentiate among practices in terms of their importance or significance. Societies could then be classified solely according to the nature of these significant practices, with variations in others disregarded. However, this requires a theoretical perspective from which the importance of diverse initiation practices can be assessed. The theoretical perspectives of interest to us in this regard are structuralism, functionalism, and a combination of the two. Let us examine how each might be applied.

Structural approaches, as we have seen, often begin with an examination of spatial organization. Applied to initiation ceremonies, this could lead to an emphasis upon such practices as the segregation of initiation compounds from the village proper, and the separation of initiates from their female relatives during the ceremonies. Thus, mental "pictures" of the ceremonies would indicate the features that should be stressed in a structural classification of societies. The analysis might then proceed to examine the relationship between such ceremonies and other social structures. Societies with initiation ceremonies, for example, have been found more likely to have legal systems in which a male kinship group incurs the liabilities of its errant individual members. Correspondingly, societies without male initiations are more likely to consider legal liabilities an individual matter.[20]

The structural classification of societies, with regard to initiation ceremonies, might therefore emphasize features according to their spatial distribution. As a specific example, all societies that place adolescent males in a separate compound might be said to have initiation ceremonies; all that do not would be said to lack such ceremonies, regardless of what other practices they might have. A structural analysis might then proceed to relate this structurally defined, or categorized, phenomenon to other variable practices, such as those associated with the legal system. The ideal result would approximate a social morphology; that is, it would describe the structure and form of societies, without regard to the functions of the classified components. Thus, one class (or species) of society might be found to have no initiation ceremonies (as structurally defined), individualized notions of legal responsibility, an educational institution of type X, a political institution of type Y, etc. Another class

[20]Cohen "The Establishment of Identity in a Social Nexus."

might be found to have initiation ceremonies, joint legal responsibility, educational institutions of type *A*, political of type *B*, and so on. Knowing the forms and structures that go together might imply functionality; that is, this information might indicate reasons for the associated practices, but to pursue these questions would shade into a structural-functional investigation.

When following a purely *functional* perspective, by contrast, studies have shown male initiation ceremonies to be most likely to occur in a society where early socialization practices foster close mother-son relationships and antagonistic father-son relationships. The risk in such societies, it is inferred, is that by identifying with their mothers, males will develop inappropriate sex identifications that will hamper their capacity as adults to perform their sex role responsibilities in an effective manner. Dramatically terminating dependence on females and the transfer of identifications from females to males thus explains, "the function of male initiation rites which accounts for the presence of these rites in some societies and the absence of them in others."[21]

This type of functional interpretation of initiation ceremonies also provides a basis for classifying societies. Specifically, in the study by Whiting and his colleagues, societies were classified as having initiations only if they possessed practices presumed to be capable of a substantial psychological impact on male youngsters. Thus, practices such as simple tatooing or wearing special clothes were judged not to constitute initiation cermonies. Societies that engaged only in such practices were classified with those societies that lacked ceremonies altogether, because such practices were assumed to be too weak or too mild to effect psychological identifications. A functional analysis would then proceed to examine the potential consequences of these practices. Remember that because such analyses occur within a system perspective, extensive ramifications are ordinarily assumed.

It is apparent that structural and functional analyses each have implications for the other. A structural analysis of the relationship between initiation ceremonies and the organization of legal liabilities suggests that one function of such ceremonies may be to unify male kin when, as a group, they share legal obligations. Similarly, the structure of such ceremonies suggests that another consequence (i.e., function) symbolically emphasizes a separation of adolescent boys from adult females.[22] Despite such implications, however, structural analyses clarify function, and functional analyses clarify structure, merely as inadvertent by-products. The two are explicitly merged in a *structural-functional* approach.

[21]Whiting, et al., "The Function of Male Initiation Ceremonies," p. 360.

[22]These different functions of initiation ceremonies have been presented in the literature as inconsistent with each other; that is, they cannot coexist. However, there is no reason why the same practice could not serve multiple functions simultaneously. Critical testing, however, has been impeded by the absence of more explicit classifications of societies such as those described here.

More on Structure and Function

Most functional theories, as previously noted, are actually structural-functional in approach, at least to some degree. They have been frequently merged not only because of their symbiotic implications, but because both perspectives emphasize a number of other conceptual features that make them highly compatible. One such shared assumption is that things are not the way they seem to be; or, not the way they are conventionally experienced by people in a society. With respect to the classical structural theories of Marx, Freud, and others, we have already noted the assumption that people see and recognize only epiphenomena, the surface feature of the underlying structure, and that these surface features are not reliable indications of the underlying structure. Marx, for example, emphasized how the proletariat misinterpreted their deprivation, taking it personally rather than attributing it to class-based exploitation that was endemic to the structure of capitalistic systems.

Similarly, classical functional theorists have assumed that customs and practices persist because of their contributions to systems, but that the actual nature of these contributions is understood only by theoretically sophisticated observers. Thus, it would be folly to accept the Thonga's or the Balinese's own interpretation of the function of their male initiation ceremonies. They would be likely to "explain" the ceremonies as tradition, as moral obligation, or as serving functions other than those inferred from the comparative analysis of these ceremonies.

The difference between stated and inferred functions has been emphasized by Merton, who calls them manifest and latent functions, respectively. The latter involve consequences that are neither recognized nor intended by participants. Thus, initiation ceremonies may change male identities even though neither initiates nor adult males are aware of this consequence. Manifest functions, by contrast, contribute to adjustment or perpetuation of a system in ways that are both intended and recognized by participants.[23]

Even though Merton's distinction between types of functions has been highly influential, it has been difficult to apply precisely. For example, what if the consequences of a practice are recognized but not intended? Should it then be more properly considered to be a manifest or a latent function? Despite such ambiguities, however, the distinction is an important one because it calls attention to two quite different meanings of the term function. Specifically, there is often confusion between motives and functions, the former involving subjective goals rather than objective consequences. The

[23]Merton, *Social Theory and Social Structure*.

manifest-latent distinction corresponds with the motive-function distinction, which will be extensively pursued in later chapters.

The proclivity of both structural and functional perspectives to conceptualize systems and to look beyond the explanations offered by participants in analyzing the systems is one reason for their frequent use together. As we have seen, clarification of structure often clarifies function and vice versa, further reinforcing their congruence.

Conclusion

Over the past three hundred years, all scientific disciplines have come to increasingly utilize system models. In this approach, the activities of a planet, a biological organ, or a social institution are all examined in relation to the larger systems of which they are a part. A functional perspective emphasizes the consequences of such activities for the larger systems. In addition, functionalism has typically assumed that entire systems are highly integrated and that they are seeking homeostasis. These assumptions and emphasis suggest the kind of questions it is most important for a theorist to ask; they also suggest what form the answers ought to take.

A major distinction among functional approaches in sociology entails the degree to which they simultaneously focus upon structure, defined as the organization of a phenomenon. Behind each of the major types of functionalism is a different conception of social structure. These varying conceptions limit the range of issues that are addressed and also circumscribe the kinds of theoretical positions that are considered appropriate. In the following chapter we will examine these varying types of functionalism.

2

varieties of functionalism

Many of the dominant perspectives in contemporary sociology are largely refinements of views that were initially formulated during what has come to be termed the *classical* period of sociological theory. Conventionally defined, this includes roughly the last half of the nineteenth century and the first two decades of the twentieth century. Comte, Marx, and Spencer, among others, were instrumental in the emergence of the distinct discipline of sociology during the earlier part of this classical era. Sociologists, including Durkheim, Pareto, and Weber then gave intellectual directions to the fledgling discipline that have continued to be highly influential.

Many of the major classical theories were explicitly functional in orientation. In some cases, essentially functionalist viewpoints were presented as theories of society, while in other cases functionalism was described as a general methodology, or a way of approaching sociological issues. The result has been a fusion of the two such that, in contemporary sociology, it is sometimes difficult to differentiate clearly between functionalism as a theory and functionalism as a methodology. It contains elements of both while fully qualifying as neither.

In this chapter we are specifically concerned with the varieties of functionalism that emerged from the classical era and their relationship to contemporary versions. This chapter is entitled "Varieties of Functionalism" in order to initially call attention to the fact that we are not addressing a monolithic perspective that has been described by a single voice. A major distinction among functional theories, and one that will be emphasized in this chapter, concerns the unit of analysis that is considered appropriate. At one extreme are theories that emphasize individual needs, which are biological in origin, and which view social institutions as functioning in order to satisfy those needs. At the other extreme are theories that emphasize the organization of society. In this instance it is the "needs of society" that are stressed, while individuals as individuals are disregarded. This difference in units of analysis corresponds with the different emphases upon structure that were discussed in the preceding chapter.[1]

Durkheim's structural-functionalism has probably been the single most influential statement in the historical development of functional theories, and it is the one we will consider first. When contemporary writers refer only to functionalism, it is most often specifically to Durkheim's statement that they are referring. Before it is explicitly discussed, however, it will be helpful to consider briefly the writings of Comte and Spencer, because Durkheim's work was both built upon theirs and sharpened by his polemical stance toward their work.

Comte and Spencer: The Background

August Comte, born into the French nobility during the period of recurrent nineteenth century revolutions, is the *father of sociology*, insofar as he first coined the name for the discipline. He likened society to an organism, viewing it as a functionally organized system in which the components were either "in harmony," or the system would change until they were.[2] Durkheim adopted this assumption of integration as well, though he questioned Comte's related assumption that the components of a society were amenable to deliberately planned change.

For Comte, an elite and learned class of men could, and should, create change. Durkheim was ambivalent, leaning more toward a belief that the incongruous parts of a system would disappear of their own accord, without any intentional intervention. This difference was related to more profound disagreements over what constituted progress and what was sociology's role as an agent of change. Comte envisioned a future society that was based upon

---

[1]For further discussion on the units of analysis in sociology, see George V. Zito, *Methodology and Meanings* (New York: Praeger, 1975).

[2]August Comte, *System of Positive Polity,* vol. 3 (New York: Burt Franklin, 1966). (Orginally published in 1876.)

more humanitarian principles, and he saw sociology as playing an instrumental role in attaining this objective. Durkheim, by contrast, emphasized a sociology that was problem-oriented, but less utopian, and he rejected Comte's assumptions of progress. In observing how the rate of suicide increased in modern civilizations, for example, Durkheim in effect asked: is this progress?[3]

The humanitarian progress envisioned by Comte was seen as resulting from a collective unity, which could emerge despite the heterogeneity and "antagonistic tendencies" of modern man. However, under these conditions, he reasoned, unity could only come about if it were imposed by an external order. While rejecting the idea that any external order would necessarily lead to progress, Durkheim elaborated upon a similar notion, viewing society as the external order at the heart of his theory. Comte's notion of an external order is therefore important to understand, but unfortunately, it was not clearly presented. He wrote poorly and was "freely" translated.[4] Nevertheless, here are some of the ways Comte characterized this external order:

1) "The subjugation of human life to this order is incontestable."[5]
2) "[It] checks the power of the selfish instincts."[6]
3) "Men have . . . been for a long time ignorant of this Order. Nevertheless we have always been subject to it; and its influence has always tended, though without our knowledge, to control . . . our actions . . . our thoughts, and even our affections."[7]

He also viewed this external order as irreducible, and as understandable through inductive reasoning:

"Meterological phenomena might result from a combination of astronomical, physical, and chemical influences. . . . But in all phenomena that are not thus reducible, we must have access to inductive reasoning. . . . Thus, the conception of an External Order is still extremely imperfect in many . . . minds because they have not verified it sufficiently in . . . the phenomena of society."[8]

In Durkheim's theory, society was the external order and he attributed to it all that Comte had attributed to the external order; namely, external restraint and regulation, and non-reducibility. Durkheim's insistence on non-

[3]Emile Durkheim, *Suicide* (New York: Free Press, 1951). (Originally published in 1893.) This movement from Comte's positivism to Durkheim's functionalism is further discussed in Alvin W. Gouldner, *The Coming Crises of Western Sociology* (New York: Basic Books, 1970), pp. 117-20.

[4]See George Simpson, *August Comte* (New York: Crowell, 1969).

[5]Comte, *System of Positive Polity,* Vol. 1, p. 17.

[6]*Ibid.*, p. 18.

[7]*Ibid.*, p. 21.

[8]*Ibid.*, p. 20.

reducibility can also be further understood as a polemic against Spencer's view that all forms of life—whether inorganic, mental, or social—are subject to the very same principles or laws. Durkheim objected to this generalization because it implied that society was not a separate and non-reducible phenomenon.

Spencer, like Darwin, was a major contributor to general evolutionary theories of the nineteenth century, and it was Spencer who first talked about survival of the fittest.[9] His first principle was the instability of the homogeneous. It maintained that differentiation among the parts of any homogeneous entity was inevitable, regardless of whether the entity was composed of biological cells, the leaves on a tree, or a group of men. External influences, such as sunlight and wind, or an outside clientele, would necessarily result in differentiation. A struggle for supremacy among the differentiated parts was considered likely to ensue, resulting in still further differentiation, and finally in the segregation of the differentiated parts.

In viewing the transformation of society as involving a movement from the homogeneous to the heterogeneous, Spencer's view strongly resembled Durkheim's later theory of the division of labor; however, Durkheim did not share Spencer's insistence on the inherent necessity of such transformations. Spencer also presented an organic biological model that was highly similar to Durkheim's. And, even though it was poorly developed, Spencer pointed to a relationship between population and social organization that was also to become an important part of Durkheim's theory of the division of labor.[10]

Despite the similarities in their perpectives, Durkheim's general stance toward Spencer, as toward Comte, was negative. This was largely due to Durkheim's view of Spencer as a reductionist; that is, of reducing society to its individual elements.[11] Indeed, Spencer did attribute social structures and processes of societal change to individuals' needs, such as desires for happiness. Individual needs and intentions had little place in Durkheim's functionalist sociology with its emphasis upon the social organism as the irreducible reality.

Societal Functionalism

In the "rules" of method, Durkheim places individuals' needs and motives outside the proper realm of sociological inquiry, identifies the social order and its elements (social facts, such as norms and rates of activity) as the appropriate subject matter, and then asks: how are these social facts to be

[9]For biographical information, see Herbert Spencer, *An Autobiography* (London: Williams and Norgate, 1904); and Ann Low-Beer, *Spencer* (New York: Macmillan, 1969).

[10]Herbert Spencer, *First Principles* (New York: Appleton, 1900), pp. 466-73.

[11]Emile Durkheim, *Rules of Sociological Method* (New York: Free Press, 1964), pp. 89-90. Durkheim viewed Comte's position in this regard as one of "ambiguous eclecticism," as stated in footnote 23, p. 122, in Durkheim's *Rules*.

explained?[12] Obviously, he answers, no individual characteristic can logically explain them because the social facts survive beyond individual lifetimes. Thus, society "greets" each generation with a world of previously developed social facts; and many of the same social facts will persist beyond their deaths. The explanation of social facts must therefore lie in previously established social facts.

At this point Durkheim seems to advocate a causal model for analyzing social facts: "No force can be engendered except by an antecedent force."[13] Functional explanations are correspondingly set apart by Durkheim, who argues that structures sometimes exist despite the fact that they serve no function ("survivals"), and that structures may remain the same while their functions change. In any case, he concludes, "the causes of its existence are, then, independent of the ends it serves."[14] This passage in Durkheim is ambiguous and subject to at least two different interpretations. On the one hand, it may mean just what it appears to mean: cause and function are two different things. On the other hand, Durkheim goes on to define "ends" in this context as referring to *individual* utilities, and he introduces the term "function" to refer to practices that satisfy the general needs of the *social* organism. Thus, rather than being intended to separate function and cause, the quotation in question may merely be a restatement of the non-reducibility perspective.

In later pages of *Rules of Sociological Method,* Durkheim oscillates, separating function and cause, and then fusing them together again. For instance, to explain a social phenomenon, he states, "we must seek separately the efficient cause which produces it and the function it fulfills"; clear enough, but then he adds, "the bond which united the cause to the effect is reciprocal . . . the . . . cause needs its effect."[15] To illustrate this latter point, Durkheim proposes that the punishment of a crime "is due to" (i.e., caused by) the collective sentiments that are offended by the crime. The punishment, he continues, is also functional for maintaining these sentiments. Unless they are periodically expressed through punishments, the sentiments themselves will diminish. If put in causal terms, this would imply that punishments cause the collective sentiments, and this is exactly the reverse of Durkheim's prior contention.

Durkheim was apparently thinking about causality in some ways, which is discrepant from contemporary modes of thought. His argument that the cause "needs" its effect, for example, runs counter to current conceptions of

[12]This discussion of Durkheim is based primarily upon the position he formulated in *Rules.* We follow it here because it is Durkheim's most explicit statement on functionalism. However, in other writings of Durkheim there is a much greater interest in, and appreciation of, the relationship between the individual and society. See Robert A. Nisbet, *The Sociology of Emile Durkheim* (New York: Oxford University Press, 1974).

[13]Durkheim, *Rules,* p. 90.

[14]*Ibid.,* p. 91.

[15]*Ibid.,* p. 95.

causality as implying one-way directionality; that is, an increase in the magnitude of the causal variable leads to an increase in the effected variable, and not vice versa.[16] In addition, Durkheim's proposed method for demonstrating causality is not tenable by contemporary standards. He argued that causality could be inferred from concomitant variation, or simple correlation. However, the sheer existence of a (zero order) relationship between two variables is not sufficient evidence. The variables may be related because they share a relationship in common with a third variable. Or they may be related because they are both caused by some other variable(s). In any event, it is extremely tenuous to observe a simple relationship and leap to a causal conclusion, and Durkheim was not sufficiently sensitive to this difficulty.

The similarities and differences between cause and function introduce a very complex topic that will be further examined in Chapter 6. For now, we may simply conclude that the relationship Durkheim saw between cause and function was not clear. However, an emphasis upon "simple," non-causal relationships is compatible with his view of society as an organic whole in which the parts are harmoniously integrated.

It was in hedging an earlier statement about the prevalence of survivals that Durkheim makes his most explicit statement of this functional model. Here he states that non-useful practices "cost effort" without reaping any benefits for the society. Survivals are therefore "parasitic" to the budget of the social organism, and no society could afford to carry very many of them. Thus, Durkheim concludes, it will typically be possible to show that social facts "combine in such a way as to put society in harmony with itself and with the environment external to it."[17]

An Italian-born contemporary of Durkheim's, Vilfredo Pareto, offered a functional perspective of society that was internally more consistent than Durkheim's, as well as methodologically more sophisticated. Specifically, Pareto argued that simple, one-sided causal models rarely were congruent with the way social phenomena operated. He supposed, for example, that there was a correlation between occupational or economic status, on the one hand, and individual's attitudes or sentiments ("residues") on the other hand. To interpret such a correlation, he insists, it is wrong to attempt "isolating economic status from other social factors, toward which . . . it stands in a relation of interdependence; and further, in envisioning a single relation of cause and effect, whereas there are many such relationships all functioning simultaneously."[18]

---

[16]For further discussion, see Herbert L. Costner, "Theory, Deduction, and Rules of Correspondence," *American Journal of Sociology,* 75 (1969).

[17]Durkheim, *Rules,* p. 97.

[18]Vilfredo Pareto, *The Mind and Society* (New York: Dover Publications, 1963), Treatise 1727. (Originally published, in Italian, in 1916.) For further discussion of Pareto's pioneering contributions to functionalism, see Joseph Lopreato, *Vilfredo Pareto* (New York: Crowell, 1965), pp. 2-22.

In a large number of other respects though, the two theorists were substantially alike.[19] Both conceptualized society as an organic social system. Also like Durkheim, Pareto conceptually emphasized structure and function and differentiated between them. The forms of religion, government, and the like, he noted, vary from society to society; but they function in highly similar ways. Finally, both utilized functional models to argue against the evolutionary theories of society, associated with Spencer and others, which had dominated sociological thinking in the late nineteenth century. To grasp the inconsistency between functional and evolutionary explanations, consider the flippers of a whale and the fins of a fish. Their evolutionary development differs, but their functions in adapting the organism to swimming are nearly identical.[20]

Despite these similarities, however, Pareto was not a societal functionalist to the same degree as Durkheim, because he violated the latter's non-reducibility axiom. "The form of a society is determined," Pareto wrote, "by . . . such elements [as] . . . race . . . sentiments . . . interests, aptitudes . . . and so on.[21] Correspondingly, in explaining social facts, such as the form of social institutions, Pareto considered such non-social prior facts as soil, climate, vegetation, and the like.

Individualistic Functionalism

Characteristic of all the perspectives that can be assigned to this category of functionalism is an emphasis upon individual needs—whether psychological or biological in origin—as providing a basic "stimulus" to social organization. Thus, social institutions, cultural values, and the like, are seen as functional responses to needs that are associated with individuals as primary units of analysis. Herbert Spencer's previously described emphasis upon the evolutionary role of individual utilities was an early statement of this orientation. However, its most complete development is historically associated with the writings of the Polish-born anthropologist, Bronislaw Malinowski.

Malinowski began with an emphasis upon innate, biologically determined needs, such as sex and hunger. These drives, whose satisfactions are

[19]Sorokin emphasizes the similarities between Durkheim and Pareto, including their functional methodologies, which he calls, "practically identical." However, this obscures the important difference noted above. Pitrim Sorokin, *Contemporary Sociological Theories* (New York: Harper, 1928), p. 45.

[20]Robert F. Spencer, "The Nature and Value of Functionalism in Anthropology," in *Functionalism in the Social Sciences,* Monograph 5, ed. Donald Martindale (Philadelphia: American Academy of Political and Social Science, 1965).

[21]Pareto, *The Mind and Society,* Treatise 2060.

imperative, are seen as setting the limits on cultural or social organizational variability. In other words, every society must make provisions for them. In addition, this biological infrastructure is also seen as resulting in certain universally human experiences, such as sexual jealousy and grief over deaths. Malinowski also viewed culture as intervening to shape the way in which these basic drives would be expressed, and as providing the modalities by which they could be satisfied. The nature of this interaction between people's basic needs and culture was illustrated by Malinowski in his analysis of the Freudian oedipal complex among the Trobriand Islanders of the South Pacific. In Western societies, he reasoned, a young male's sexual jealousies and resentments are directed at his father because they are shaped in that direction by typical family structures. Among the Trobrianders, socialization authority was exerted more by a mother's brother (i.e., maternal uncle) than by a father. Correspondingly, Malinowski reported that oedipal-like resentments of young Trobriand males were primarily directed at maternal uncles, and maternal cousins rather than mothers were the objects of prohibited sexual desires.[22]

Malinowski, as previously noted, viewed culture not only as the shaper of human drives, but also as the provider of functional responses to these drives. People's basic institutions were seen as collective responses to these needs. The form such institutions might take was seen as highly variable, as variable as culture itself; yet, each institution in its own way addressed highly uniform needs. For example, the mortal nature of man assures that there will be recurrent deaths, accompanied by specific sorrows. Cultures respond by providing institutionalized means for the expression and relief of grief, though in very different kinds of ways. Again among the Trobrianders, Malinowski described an example that was far different from conventional Western practices. Elaborate and extended Trobriand funeral-mourning services begin with the washing and decorating of a corpse. Continuing for up to several years, the corpse is buried and exhumed, with its bones distributed among relatives who wear them as symbols of respect. Special obligations, also lasting as long as several years, fall to certain survivors. After the death of a husband, for example, a widow is confined to a "cage" in her house, and is not permitted to feed herself or to speak in a voice above a whisper for as long as two years.

From culture to culture, Malinowski concluded, mourning rituals vary. They all function similarly, however, in providing institutionalized means for expressing grief. Similarly, religion, magic, and other more or less universal institutions were all seen as functioning to satisfy basic human needs in variable cultural contexts. Each institution has a "charter" to perform, Malinowski claimed, and he correspondingly rejected the idea of survivals. If the basic need being served by the institution was difficult to identify, then

[22]Bronislaw Malinowski, *Sex and Repression in Savage Society* (New York: Harcourt Brace, 1927).

Malinowski shifted levels of analysis and viewed continuation of the practice as contributing to social cohesion. At least this function would always be served, an assumption that logically precluded the possibility of survivals.

Also at a non-individualistic level of analysis, Malinowski argued that social institutions themselves must theoretically be interrelated. This assertion followed from his view of cultural systems as involving totally integrated ways of life. In describing his Trobriand Island field work, for example, he stated that the really difficult task was not making observations or getting facts, but trying to, systematize them into an organic whole.''[23] In line with this view of the total interrelatedness of cultural systems, Malinowski stressed that tampering with any practice in a society, regardless of its apparent insignificance, would likely have profound repercussions throughout the system. Any tampering, in other words, would create the risk of creating unanticipated, and often undesirable, changes throughout all the other related institutions.

The extensiveness of the degree of interrelationship among the institutions in societies of the type analyzed by Malinowski is illustrated by Sharp's analysis of the Yir Yoront. This tribe of Australian aborigines had maintained an "essentially paleolithic society" until the middle of the twentieth century, and many of their vital traditional institutions involved stone axes.[24] These axes had historically been owned only by men, and typically by older men. Women and children needed to use them regularly, to cut firewood, repair huts, etc. The superordinate status of the males was symbolically reaffirmed every time someone in a subordinate status borrowed their axe. Thus, men's exclusive ownership of this vital implement kept women and not-yet-initiated males "in their place." In addition, the men obtained the stone for the axe from distant trading partners, an exchange that only they were permitted to make. Their exchanges occurred during the annual dry season, which was also the time when initiation rites and other ceremonies occurred.

The demise of the traditional Yir Yoront culture was brought about by an Anglican mission station, which in order to raise native living standards indiscriminately distributed steel axes to men, women, and children. This new pattern of ownership directly disrupted traditional sex- and age-role relationships, and ultimately destroyed the historic trading partnerships and ritual ceremonies, including initiation, which were associated with the exchanges. These changes made the Yir Yoront still more vulnerable to other European influences that were sweeping across the peninsula. The traditional culture died, and though the people did not, they were left confused and bewildered because of the changes. Thus, so integrated was the cultural system, that one

[23]Bronislaw Malinowski, *Coral Gardens and Their Magic,* Volume 1 (New York: Harcourt Brace, 1952), p. 322.
[24]Lauriston Sharp, "Steel Axes for Stone Age Australians," in *Human Problems in Technological Change,* ed. Edward H. Spicer, (New York: Russell Sage, 1952).

change—the innovation of the steel axe—led to profound repercussions throughout the system.[25]

For Sharp, as well as for a whole generation of British social anthropologists, Malinowski's detailed and insightful field work became a research model. They were also strongly influenced by Malinowski's functional approach, which was solidly opposed to an evolutionary perspective that had been equally dominant in both anthropology and sociology. At the same time, however, they were influenced by the anti-reductionistic functionalism of Durkheim and others. As they moved in this direction, an increasing emphasis was placed upon the idea that kinship systems provide the basic structure upon which societies and cultures are built. Raymond Firth's writings are particularly illustrative.

Firth's best known study was an analysis of the Tikopia who, like the Triobrianders, lived in the South Pacific islands. He emphasized that among the Tikopia, as in other primitive societies, kinship links provided the skeleton that gave form to the entire society. Patterns of economic cooperation, religious ritual, and other institutionalized activities were seen as erected upon kinship, which made analysis of kinship analogous to "investigating . . . the physiology of the society."[26] This clearly introduced a structural concern with institutional form, and with the integration of institutions. This structural emphasis was introduced by Firth without explicit rejection of Malinowski's view that institutions were a functional response to individual needs. However, this structural emphasis led Firth to question the needs of the social structure itself, in a very Durkheimian manner. In analyzing Tikopia kinship groupings and patterns of descent, for example, Firth asserts that there are "certain fundamental conditions to which a society . . . must conform if it is to maintain its existence."[27] As examples of such requirements he notes economic cooperation, kinship stability, and social control norms within the society. Firth also comments that socially necessary customs often produce strain and friction among the individuals whose behavior they direct. Thus, running counter to Malinowski's position, there is the view that the needs of individuals and the needs of society may be imcompatible.

Interpersonal Functionalism

The most direct attack on Malinowski's individualistic functionalism was offered by A. R. Radcliffe-Brown, another leading figure in British social anthropology. In his most insightful essays, Radcliffe-Brown focused upon

[25]Sharp notes that steel axes were actually one of many simultaneous European innovations, but clearly the most important one.

[26]Raymond Firth, *We, the Tikopia* (London: Allen and Unwin, 1936), p. 575. (The preface to this book was written by Malinowski.)

[27]*Ibid.*, p. 576.

the patterning of social relationships. This involved emphasis upon interpersonal interactions, a unit of analysis lying somewhere between Durkheim's and Malinowski's, but clearly closer to Durkheim's theory. More specifically, he was interested in the functions of special strain-accommodating mechanisms that were built into the structure of ongoing social relationships. Examples include such patterned interpersonal practices as joking, gift-giving, avoidance, and privileged familiarity. The most important question to answer according to Radcliffe-Brown, is: how do these patterned relationships function to connect individuals into an integrated whole?[28]

In analyzing relationships of this type, he began from the premise that numerous relationships must be attended to by only minimal amounts of conflict if the social order is to persist; but these very same relationships may involve built-in strains. The relationship between a man and his mother-in-law, especially in many primitive societies, is representative. Given the important role played by extended kinship groups in economic and subsistence activities, a reasonably high degree of solidarity among extended kin is required. This necessitates a minimum amount of conflict, but men and their mothers-in-law may be structurally locked-in to a conflicting situation. They may, for example, have competing demands upon the services or loyalties of the wife-daughter, who is the connecting force between these individuals.

A normative pattern in which a man is expected to avoid interacting with his mother-in-law is one solution Radcliffe-Brown observed for minimizing overt tensions. A pattern of expected joking, in which no offense is to be taken at the teasing, is another way. Both accomplish the same objective, though in different ways.[29] In terminology developed later, avoidance and joking would be termed "functional equivalents."

Radcliffe-Brown also raised the interesting question of how to know whether the above is really the function. His answer, like Durkheim's, emphasized comparative analysis of the conditions under which a phenomenon occurs. Only then could its function be deduced. Correspondingly, and again like Durkheim, he emphatically rejected the stated motives of participants as constituting the explanation. Thus, the important functions to understand were the latent ones; as defined in Chapter 1 the functions that were neither recognized nor intended. By definition, therefore, participants' views had to be treated lightly in favor of an investigator's own deductions from comparative study. In this context, Radcliffe-Brown reports one society's interpretation of their joking relationships as entailing "the purification of the liver." Facetiously he asks, should we accept their assessment? Similarly, in an analysis of rituals, he describes Australian totemic rites, which are seen by the natives as

[28]A. R. Radcliffe-Brown, "On the Concept of Function in Social Science," *American Anthropologist,* 37 (1935). Reprinted in a collection of essays, A. R. Radcliffe-Brown, ed., *Structure and Function in Primitive Society* (New York: Free Press, 1955).

[29]Radcliffe-Brown, "On Joking Relationships," in *Structure and Function in Primitive Society.*

serving to maintain some species of plant or animal life (depending upon the specific ritual). Rhetorically he again asks, should the participants' motives be accepted? His serious answer is that stated purposes are highly relevant to a consideration of psychological functions, but only tangentially relevant to social functions.[30] This statement almost paraphrased Durkheim's point, namely, that there were as many stated reasons for suicides as there were suicides, but none of the stated reasons contributed very much to a sociological explanation of suicide as a social fact.

The very core of Radcliffe-Brown's disagreement with Malinowski was expressed in their differing views on the nature and function of magic. In his Trobriand Island observations, Malinowski noted that open-sea fishing was accompanied by a great deal of ritual. As the boats moved further from the shore, Trobriand fishermen increasingly utilized charms and amulets. When they fished at the inner lagoons, however, there was no comparable use of magic. The two types of fishing differed, Malinowski contended, in the degrees of danger and uncertainty that were involved. Open-sea ventures were more precarious and the likelihood of success was more problematic. Given the people's ignorance of the efficacy of magic, Malinowski viewed the Trobrianders as resorting to supernatural practices in response to their feelings of danger and uncertainty. Note that this interpretation is perfectly consistent with Malinowski's general view of social institutions as a functional response to individual needs.

Radcliffe-Brown, as might be anticipated, preferred a different explanation. Moving back a step, he insisted that is was society that defines the uncertainties in life and dictates the appropriate response; whether religious supplication, magical invocation, or the like. In effect, society would not be possible unless people learned what to feel, and when and how to express their feelings. Births and deaths, for example, are significant events for a *society,* entailing the entrance and exit of its "members." This would be true for society, regardless of individual sentiments pertaining to births and deaths. In order for individuals to be linked together, however, they must be taught to have the appropriate feelings. This is viewed by Radcliffe-Brown as the "essential function and ultimate reason" for the existence of magic or religion. These rites, he states, "exist and persist because they are part of the mechanism by which an orderly society maintains itself in existence, serving as they do to establish certain fundamental social values."[31]

Radcliffe-Brown's position here is, again, highly congruent with Durkheim's. In his analysis of religion in primitive Australian tribes, Durkheim argued that there was a very arbitrary differentiation made among two classes of ideas and objects in a society. Some were considered sacred (special

[30]Radcliffe-Brown, "Taboo," in *Structure and Function in Primitive Society.*
[31]*Ibid.,* p. 152.

and set apart) while others were considered profane (ordinary and everyday). They were differentiated from each other not by their inherent qualities, but by the sentiments that people were taught to have regarding each. The sacred objects, Durkheim continued, correspond with activities that are of special importance to the society. Therefore, the objects deemed sacred by religious or magical beliefs are actually symbols of the society, and elevating them to a sacred status works to reinforce collective solidarity.[32]

Thus, Radcliffe-Brown and Durkheim both rejected the idea that individual needs have an independent autonomy capable of determining the nature of social institutions. Anticipating this later disagreement between Radcliffe-Brown and Malinowski, Durkheim wrote: "For those who profess the complete autonomy of the individual, man's dignity is diminished whenever he is made to feel that he is not completely self-determinant . . . however, most of our ideas and our tendencies are not developed by ourselves but come to us from without . . . by imposing themselves upon us."[33] It was, of course, society that Durkheim viewed as the imposing external force.

Empirical Applications

The individualistic and the societal forms of functionalism have both stimulated research by ensuing sociologists. The results of such studies have generally been supportive of, or consistent with, the perspectives that stimulated them. It is debatable whether abstract perspectives such as these are amenable to direct empirical assessment. However, even where assessment is not possible, research studies often have a pedagogical value, which is: helping to clarify the more abstract perspectives that guided their formulation. With this intent in mind, let us briefly consider some representative studies.

Following Malinowski, Lionel Lewis has examined the empirical question of whether the utilization of magical or religious practices is associated with individual experiences of uncertainty and danger.[34] He specifically studied the practices of slightly over 100 women whose children were hospitalized with serious illnesses. From questionnaires, Lewis measured the degree to which they believed their children's illnesses to be dangerous, and their amount of knowledge concerning the illness and its treatment. Among the mothers who were low in knowledge, the more they perceived the illness as dangerous the greater was their reliance on magical charms, superstitions,

[32]Emile Durkheim, *Elementary Forms of the Religious Life* (New York: Macmillan, 1915).

[33]Durkheim, *Rules*.

[34]Lionel S. Lewis, "Knowledge, Danger, Certainty and the Theory of Magic," *American Journal of Sociology,* 68 (1963).

etc. On the other hand, the greater their feelings of uncertainty, the more they prayed, read bibles, and engaged in other forms of religious supplication.

Lewis did not continue the investigation to see whether the mothers' magical and religious practices actually lessened their fears and uncertainties. Assuming that they would, however, Lewis's study clearly illustrates Malinowski's emphasis upon the way in which social practices function to meet individual needs.

Following Durkheim and Radcliffe-Brown, Guy Swanson examined the relationship between social organization and the forms of religious expression in a sample of about fifty primitive societies.[35] He reports that in societies where there is a relatively large number of sovereign groups who are responsible for certain activities, religious expression is typically organized around the belief in a single high God. By contrast, in societies where there are few such groups, religious expression typically involves numerous lower deities whose powers are more limited than those of a single high God.

The difference, according to Swanson, lies in the need for total unification in a society, which is most effectively provided for by the belief in one God. Specifically, the greater the number of sovereign groups, the more problematic is solidarity or cohesion of the entire society. In such societies, therefore, a single high God is more functional for societal unification. Swanson did not extend this analysis to determine whether fragmented societies with a high God are indeed more effectively unified than those with numerous deities. The assumption that they are, however, is what makes the observed relationship theoretically meaningful. In any event, Swanson's study clearly illustrates Durkheim's view that social facts (such as forms of religious expression) function to satisfy non-reducible needs of the social organism.

An interesting question arises at this point, particularly when we consider the implications of research such as that discussed in the preceding pages. Is it possible for this research, or any research for that matter, to support agruments concerning the "primacy" of individuals or of society in functional analyses? To a limited degree it is, in that any perspective that fails to generate researchable hypotheses would likely be dismissed as barren or fruitless. Both the individualistic and societal views, however, have stimulated research, as we have seen.

It is also important to recognize that arguments concerning the primacy of individuals or of society raise issues that are both metaphysical and epistemological, as well as empirical. As such, their validity is only partially demonstrable. For example, to view individual needs as socially determined, and therefore as lacking the autonomy required to serve as bases for social

[35]Guy E. Swanson, *The Birth of the Gods* (Ann Arbor: University of Michigan, 1960). Swanson extends and clarifies this argument in "Monotheism, Materialism and Collective Purpose," *American Journal of Sociology,* 80 (1975).

institutions, is too abstract an assertion to be completely assessed empirically. Further, it is intended only in part to be a testable assertion. In large measure it is also intended to convey a philosophical spirit of inquiry; that is, to define and delimit the boundaries of sociological theory and sociological investigation. This dimension of the assertion is not amenable to direct empirical assessment.

Parsons's Synthesis

One way out of the either/or dilemma concerning individual or societal functions is to deny that it is a matter of choice, and opt for both. This was the route taken by Talcott Parsons, whose functionalist perspective became dominant in the mid-twentieth century. (Some sociologists would argue that it is still dominant.) Parsons was influenced by a number of prior functionalist theories, notably those of Durkheim and Pareto; he was also influenced by Freud and Weber as well as by a number of functionally-oriented culture and personality theorists in social anthropology. In effect, Parsons attempted to throw a blanket over all of these diverse perspectives and unify them into one coherent theory of action. As sociologists, Parsons stated, we are primarily interested in the social system, which he roughly equated with society and whose dominant function he saw as integration. However, cultural and personality systems were seen as impinging so directly on the social system that their influences could be separated out only in arbitrary conceptualizations.

Parsons's most fundamental concern is with how social order is possible; specifically, the Hobbesian question of what prevents continuous "war" among persons in a society. In large measure he sees the answer amid the interpenetration of the cultural, social, and personality systems. Correspondingly, he describes societies in which cultural values are institutionalized in the social system, and whose norms and rules are, in turn, internalized in personality systems.[36] Social order, then, is possible because people will comply with expectations without experiencing conflict. In other words, the social rules "possess their harmonious character by virtue of their derivation . . . from common value orientations which are the same for all."[37] They are therefore viewed as legitimate; but individuals also comply

[36]All three were also viewed as subsystems in his general theory of action. Talcott Parsons, et al., *Theories of Society* (New York: Free Press, 1961). For a more brief and clear presentation see *The System of Modern Societies* (Englewood Cliffs, N. J.: Prentice-Hall, 1971).

[37]Talcott Parsons and Edward A. Shils, *Toward a General Theory of Action* (Cambridge: Harvard University, 1951); and, Parsons and Shils, *Toward a General Theory of Action* (New York: Harper and Row, 1962), p. 194.

because of the overlap or integration between their needs and the culturally derived rules of the social system. Thus, their internalized values provide a personal motivation for complying with legitimate role expectations.

Parsons proceeded to establish conceptual models that could simultaneously be applied to the different units of analysis; i.e., personal, social, and cultural. Among his best known efforts in this area are the pattern variables, a classificatory scheme that describes the alternatives in any system of action. It includes five dichotomous variables, which according to Parsons, exhausts the possibilities of actual behavior, normative expectations, or cultural values. The five variables, and their attendant choices, are the following:[38]

1) Affectivity or neutrality—to seek or to control gratifications
2) Self or collectivity—to give primacy to private or to shared interests
3) Universalism or particularism—to stress general or personal standards
4) Ascription or achievement—to emphasize attributes or performances
5) Specificity or diffuseness—to respond restrictively or to generalize

Each of these pattern variables are seen as applicable to all units of analysis. At the personal level they refer to need dispositions; at a social-system level they refer to role expectations; at a cultural level they refer to value orientations. By viewing motives, orientations, and values as arranged along the same axes, the pattern variables present a model that greatly facilitates inter-unit, or inter-system, comparisons.

While Parsons did not expect the ideal type condition of perfect congruence among the systems to be completely realized in any actual society, a substantial degree of its attainment was perceived as both inevitable, and as the answer to Hobbes's question. Institutionalization and internalization were thereby formulated as functional prerequisites of society, implying that to some degree: 1) they *would* be present in any society, and 2) they would *have* to be present for any society to continue. The conception of functional prerequisites for a society had been building toward just such a formalized statement during several decades of functionalist writings. Firth, for example, had viewed familial control over sexual drives, among other institutionalized practices, as constituting necessary "fundamental conditions." Radcliffe-Brown had suggested that Durkheim's concept of function be replaced with the notion of the social organism's "necessary conditions of existence." Then, about 1950, Parsons became one of many theorists to offer a detailed list of such prerequisites.[39] In fact, much of Parsons's entire paradigm is stated in

---

[38]*Ibid.*, pp. 47-109.

[39]See, for example, David F. Aberle, et al., "The Functional Prerequisites of a Society," *Ethics*, 60 (1950). Many of the functional prerequisite arguments were directed primarily at small, relatively uncomplex, and primitive societies. For Parsons, who was also addressing contemporary societies, this led to a concern with special mechanisms that might be operating only in complex societies. With respect to the economy, for example, see Talcott Parsons and Neil J. Smelser, *Economy and Society* (New York: Free Press, 1956).

terms of functional necessities. With respect to socialization, for example, he states that it *must* result in the acquisition of orientations that are congruent with the roles people *must* play. Therefore, "there must be mechanisms of social control operating on the parents as well," he continues, in order to prevent "misfiring of the process of socialization."[40]

Societies, as conjured up by Parsons's description of social systems, are characterized by extreme stability. When socialization processes are operating as they are supposed to and they are resulting in internalization, and when social institutions embody widely shared cultural values, Parsons questions, From what source could pressures toward change or conflict emanate? Only from outside of the system, he answers, entailing, for example, a change in the physical environment.[41] Even if a proclivity toward change should somehow be instigated, however, vested interests are seen as mobilized to resist the change. These vested interests are, in effect, system loyalties. They are products of internalization and institutionalization, which direct the people's motives and values—and hence loyalties—toward the system.

From Parsons's perspective, deviance and conflict necessarily overlap with social change. In the ideal model of the social system, deviance, conflict, and change all emanate from outside the system, and all are viewed as the result of socialization deficiencies; we will discuss this further in the following chapters. For now we may conclude that the social system, as outlined by Parsons, is characterized by stability and equilibrium. More specifically, society is seen as a boundary-maintaining system that is able to perpetuate a stable equilibrium. A high degree of integration among the components of the system (e.g., institutions) is also assumed, and change anywhere in the system is correspondingly expected to have unanticipatable repercussions throughout the system.

Conclusion

In this chapter we have examined the historical development of several types of functional theories. An individualistic emphasis, associated with Malinowski, poses as its central question, How do social institutions function to satisfy individual needs? A societal emphasis, by contrast, treats the social organism as a sui generis, viewing individual needs as socially determined. Its central issue involves the way in which social institutions function to meet the needs of the social system, or of the non-reducible collectivity. Parsons's synthesis merged the two, emphasizing the interpenetration of the social and the personality, as well as the cultural, systems.

[40]Talcott Parsons, *The Social System* (New York: Free Press, 1950), p. 504.

[41]In the final chapter of *The Social System,* Parsons claims to have deliberately exaggerated a model of stability to provide a base line from which to infer change.

While the different theories that have been discussed are not always in perfect agreement in their conceptualizations of society as a functionally organized system, they do tend to be compatible on a number of points. All tend to view societies as highly integrated entities, and they share a "discomfort" with social change because of a perceived inability to predict beforehand the ultimate implications of even seemingly minor changes. In addition, the theories tend to focus upon a supra-individual unit of analysis and to emphasize the functional necessities of its continued operation. Hobbes's question, in one form or another, appears to be in the back of almost everyone's mind. Malinowski is an obvious exception here; however, it should be noted that while emphasizing individual needs, Malinowski also viewed a cultural system as an integrated whole. Finally, all of the theories tend to stress latent functions; that is, those consequences that are neither intended nor recognized and, as such, are known best by the term "outsiders."

These generally shared characteristics of functional theories have stimulated several recent decades of criticism and debate. More specifically, these theories have been accused of ideologically supporting the status quo, of being conceptually unable to analyze social change or social conflict, of being unduly wedded to a model of institutional integration, and of emasculating individuality and moral autonomy. With the clash of intellectual sabers in the background, we turn to these criticisms and their rejoinders in Chapter 3.

3

criticism and debate

Functionalism has probably provided the largest and most significant battleground in the recent history of sociology. Attacks have been frequent and extensive, while defense and counterattack have been surprisingly sparing. An observer keeping score could easily be led to conclude that functionalism had been routed; but this is not the case. The relative absence of defense is due partly to the historic centrality of functionalism in sociological theory. As a result, sociologists are unsure about what would be left of sociological theory if it were expurgated of all functionalist content. In other words, while much of the criticism has been well taken, no one fully understands what to do about it, short of starting over. Sociologists have been understandably reluctant to do this, and the danger of going too far is apparent. Thus, metaphorically, functionalism has ambled along like a giant elephant, ignoring the stings of gnats, even as the swarm of attackers takes its toll. Another reason for the relative absence of defense is that much of the criticism has been directed at a straw man, an exaggerated version of functionalism that nobody advocated in the first place. Therefore, it has no defenders. In addition, however, more recent functionalist writings have "quietly" responded to prior criticisms, expanding their treatment of previously slighted issues.

While an observer might be led, for the above reasons, to misperceive the score in the arena, the importance of functionalism to sociological theory dictates that serious criticisms of functionalism must have profound effects upon the intellectual life of the discipline. In this book, such criticism is divided into two types: that directed at ambiguities and distortions inherent to the perspective, and that which is directed at specific substantive applications of the perspective. It is theoretically important to distinguish between these two types of criticism even though they are not completely discrete. Thus, while this chapter focuses upon the general criticism and debate, many of the same issues will reappear in various forms in the substantive chapters that follow.

Ambiguity of the Concept

From the inception of functionalism there has been continuing confusion concerning the meaning of the term, "function." Even among those committed to the perspective there has been confusion and disagreement. Radcliffe-Brown, for example, viewed Durkheim's conception of function in relation to the needs of the social organism as ambiguous and teleological.[1] It is hardly surprising, therefore, that critics of the perspective would emphasize the ambiguity of the concept.

Some of the confusion surrounding the concept stems from the fact that it can have numerous referents. Thus, functional is used to refer to a purely quantitative relationship in the sense that X is a function of (i.e., varies with) $Y;$ it is used in reference to tasks and roles in the sense that the incumbent of a position is called a functionary; and it is utilized to refer both to subjective motives and objective consequences.[2] The list could go on and on. If these varied usages were completely different from each other, they would produce little confusion; unfortunately though, they are not. Thus, the term function is ambiguous because it is used inconsistently and without clearly identified referents.

If we differentiate among the types of functionalism, we can clear up this ambiguity, at least to a point. However, we will see then, other problems emerge. In societal functionalism, the term function generally refers to the

---

[1] A. R. Radcliffe-Brown, "On the Concept of Function in Social Science," in A. R. Radcliffe-Brown, ed., *Structure and Function in Primitive Society* (New York: Free Press, 1955).

[2] Robert K. Merton, *Social Theory and Social Structure* (New York: Free Press, 1949), Chap. One. See also, Ernest Nagel, *The Structure of Science* (New York: Harcourt Brace, 1961).

contribution of an element (such as a custom or institution) to the larger system of which it is a part. The key word now is contribution, which leads to the question, What is it that elements do contribute to a system? The typical answer, offered implicitly or explicitly in functionalist works, is that they contribute to the society's maintenance and/or its perpetuation. This assumption, Spencer declares, "is frankly teleological. It argues that the purpose of society is its own preservation."[3] Thus, society is viewed as having an ultimate or final goal, a teleological position that modern sciences have tried to avoid because it is not amenable to empirical assessment. Note that with a corresponding emphasis upon latent functions, societal functionalism ultimately assumes that society is motivated, as an individual might be, to persevere. Logically, the very notion of a societal motive seems to involve reification, or reductionism in reverse.

The inability to document societal motives raises the problem of functionalism's susceptibility to tautological explanations; that is, to explanations that cannot be disproved because they define things in relation to themselves. Thus, a practice is said to persist because it contributes to the maintenance of a society. Its persistence, therefore, is taken to prove that the society's primary motivation is to maintain itself. On the other hand, if the practice was discontinued, it would be taken to indicate that the society's maintenance was not dependent upon the practice's continuation. Note here that such theorizing cannot be disproved, which is the first clue that it involves tautological reasoning. It seems to offer descriptive statements about how societies operate, when in fact the statements are closer to definitions. Descriptive statements can often be molded into testable hypotheses; definitions cannot.

It is often claimed that theories involving tautological reasoning are logically and scientifically worthless as a result. This is probably an overly harsh judgment, though no one would argue that tautologies are an asset to a theory! Kaplan argues convincingly that they are not a fatal flaw, if we distinguish between logic-in-use and reconstructed logic.[4] This distinction rests upon a prior distinction made between two views of science, one of which stresses an inventory of facts and a second that emphasizes the investigative process in which such facts are generated. A logic-in-use is associated with the latter. More specifically, investigation as a process often involves a number of poorly articulated assumptions—a "cognitive style"—which may be more or less logical. The major criterion of an investigatory conception, however, is usefulness. Newton and his followers, for example, made excellent use of the calculus in physics even though its foundations were not explicitly and logically formulated until two hundred years later.

[3]Robert Spencer, "The Nature and Value of Functionalism in Anthropology," in *Functionalism in the Social Sciences,* ed. Don Martindale (Philadelphia: American Academy of Political and Social Science, 1965).

[4]Abraham Kaplan, *The Conduct of Inquiry* (San Francisco: Chandler, 1964).

This history of calculus illustrates the general point that a logic-in-use often precedes a reconstructed logic. However, the crucial question is whether the logic-in-use is helpful in clarifying intellectual problems. Again and again, Kaplan emphasizes, the major issue is usefulness and not form. Thus, even tautologies can be useful. In this regard he quotes Toulmin's disdain for the long debated question of whether Newton formulated laws or definitions. "Laws themselves do not do anything; it is we who do things with them; the important issue is, "What kinds of things can we do with their help."[5]

In sum, functionalism can be dismissed as an ambiguous, teleological, and tautological perspective whose assumptions violate logic and defy empirical assessment. On the other hand, despite such serious problems, functionalism may present a logic-in-use, or cognitive type, that is extremely useful in analyzing sociological issues. In addition, the teleological and tautological problems become less serious when it is recognized that most concrete functional studies focus upon components of systems rather than total systems. Thus, the question of whether society has an ultimate purpose can be relegated to the background. The societal functionalists in particular, however, have traditionally been sloppy about delineating functions. By overly assuming consensus and integration, they have not seen the importance of specifying functions in relation to units. In other words, in describing contributions they have often failed to ask, "Who benefits?"

Consensus, Stability, and Integration

Observations made in primitive societies were an important source of data in the formulation of many new functional theories, and they provided an important testing ground for many of the previously developed theories. The perspectives of Durkheim, Malinowski, Radcliffe-Brown, and many of the other influential functionalists were all shaped by fieldwork in remote islands, isolated villages, and jungles. The societies they encountered all tended to have simple social organizations by comparison to modern industrial societies; for example, they possessed fewer specialized groups and less elaborate stratification hierarchies. As a result, integration based upon consensus and homogeneity may have been more typical of these primitive societies than of modern societies. In addition, all the societies studied tended to be characterized by relative isolation. Integration and consensus may also have been promoted by these societies' lack of routine exposure to external influences.

The functional theories that were formulated were not confined in ap-

[5]S. E. Toulmin, *Philosophy of Science,* quoted in Kaplan, *The Conduct of Inquiry,* p. 101.

plicability to these primitive societies, but these societies did provide the model for the conceptualization of all societies. Thus, the functionalist model of society, including modern societies, has been repeatedly criticized for exaggerating consensus, stability, and integration to the point of virtual disregard of conflict, change, and disorder.

It is probable that the primitive societies studied in the early part of this century actually did differ from modern societies in the ways indicated above. In addition though, even their consensus, stability, and integration may have been exaggerated because it was Hobbes's question—How is social order possible?—that lurked in the back of the early functionalists' minds. Malinowski, for example, stressed how Trobriand exchange systems were characterized by stability and consensus. Later analyses, however, have paid more attention to the conflict that was generated by the system, and the way in which ambitious individuals could attain mobility via the exchange system.[6] Both sets of processes were apparently characteristic of the Trobrianders, although Malinowski emphasized only one side of the issue.

While all the major functionalists tended to stress consensus and integration, they did not all emphasize each similarly, or view their relationship to each other in the same way. Parsons's view, while not typical of all functionalists, has been widely criticized as the functionalist prototype. In Parsons's theory, consensus on values and ideas is highly emphasized and elevated to a crucial role. The entire social system, or at least its stability, is seen as resting heavily upon shared values, even though society is also seen as having structure and organization. In Durkheim's view, by contrast, there is a greater separation of values and ideas from social structure. Thus, Durkheim viewed society as able to direct or coerce consciousness. In sum, for Parsons a great deal of consensus is both typical and imperative. For other functionalists, the consensual requirements of a society are less clear. What, for example, is a typically standard deviation in the society surrounding Durkheim's hypothetically average man? Radcliffe-Brown, to further clarify, was explicitly interested in conflict and hostility as recurrent features of social life. He viewed divergent interests as built into the structure of extended kinship groupings, for example, and viewed patterned relationships as mechanisms for stabilizing the strife that ensued. Finally, working from an explicitly functionalist perspective, sociologists have analyzed conflict in the same manner as any other practice. Thus, Coser proposed that conflict often contributes to the maintenance of social relationships by helping to revitalize norms, serving to release tension, and so on.[7]

To keep the issue in perspective, however, it must be noted that functionalist excursions into the realm of conflict, while not non-existent, are not

[6]See, for example, Cyril S. Belshaw, *Traditional Exchange and Modern Markets* (Englewood Cliffs, N.J.: Prentice-Hall, 1965).

[7]Lewis A. Coser, *The Functions of Social Conflict* (New York: Free Press, 1956).

commonplace. And as previously indicated, critics of the functionalist emphasis upon consensus have tended to specifically address Parsons's view in which conflict was virtually disregarded except for occasional treatments of conflict as constituting a sporadic pathology. An important reason for this relative neglect of conflict is attributed, by Gouldner, to the historical emphasis the functionalists have given to the concept of reciprocity. He begins with the problematic stature of survivals in functional theories.

Gouldner's critique

Survivals can be viewed as instances of unequal reciprocity and as practices which by sheer virtue of their persistence take from "the budget of the social organism," in Durkheim's terms, while contributing nothing. Thus, Durkheim concluded, no society could afford to have very many of them. Functionality was a "working hypothesis" for Radcliffe-Brown: "The hypothesis does not require the dogmatic assertion that everything in the life of every community has a function. It only requires the assumption that it *may* have one."[8] Thus, the rejection of survivals by the functionalists was more partial and more hesitant than Gouldner notes.

Gouldner is correct, however, in noting that the functionalists did not fully allow for survivals, or vigorously pursue possible instances of them. Their rejection of survivals, Gouldner contends, is consistent with the functionalists' emphasis upon reciprocity among the components of a system; a reciprocity that results in an integrated and "harmonious" whole. Had the functionalists pursued the notion of survival, it could have conceptually opened up the entire category of unequal exchanges that could, in turn, have led to a greater interest in conflict. The Marxian notion of exploitation, for example, involves precisely a situation in which unequal exchange is built into the fabric of a society. An interest in unequal exchange could also lead to an emphasis upon coercion as an explanation for the persistence of the exploitive element. This too could have made structured conflicts an apparent issue for the functionalists to confront.[9]

Focusing specifically upon Parsons, Gouldner also argues that an emphasis upon equal reciprocity in interpersonal relations also led to an obscuring of interest in inherent conflicts. The actors in Parsons's scheme comply with each others' expectations because they want to (institutionalization) and because structurally the expectations associated with the roles of each are the obligations of the other (complementarity). This notion of a stable system of

---

[8]Radcliffe-Brown, *Structure and Function in Primitive Society,* p. 184. (Italics in original.)

[9]Alvin W. Gouldner, "The Norm of Reciprocity," *American Sociological Review,* 25 (1960). The question of unequal reciprocity is pursued in Alvin W. Gouldner and Richard Peterson, *Technology and the Moral Order* (Indianapolis: Bobbs-Merrill, 1962).

reciprocated gratifications leads to a virtual disregard of situations in which actors feel constrained to try to defend or enlarge their rights vis-à-vis another's obligations, or need to resort to coercion in order to force another to make good on a past debt. In short, Gouldner concludes, the emphases upon institutionalization and complementarity direct attention away from all potential conflicts.

Further, even the view of roles and role relationships involved in Parsons's concept of complementarity is presented in a way that inhibits conflict from entering the picture. Parsons stresses "manifest" roles and identities, which Gouldner defines as those generally considered to be relevant in any social context. In a classroom, for example, the roles of students and professors are manifest. People's expectations about others' behavior are structured by their views of these relevant characteristics of the roles. However, Gouldner contends, the actors also occupy latent roles; that is, there is another set of expectations structured about their sex, their race, etc. While these latent roles would generally be viewed by the participants as irrelevant or inappropriate to consider, they nevertheless structure expectations, sometimes in contradictory ways. Historic examples of such contradictions include female engineers and black physicians.[10] In an occupational context, their professional roles are manifest; but persons interacting with them may have been influenced by their latent (race or sex) roles, and expect different behavior from them as a result. Parsons's exclusive focus upon manifest roles is seen by Gouldner as obscuring the way latent roles can provide a basis for the emergence of conflict, as witnessed in the feminist and black movements. In addition, incongruent manifest and latent roles may create strains upon the roles' incumbent; strains that may interfere with the incumbent's ability to comply with manifest role expectations, and hence lead to conflict.[11]

the beneficiary

Merton's critique of functionalism, which stresses the problem of specifying units, further clarifies the relationship between the functionalists' assumption of social integration and their relative disregard of conflict. The assumption that societies are functionally integrated was held by all the functionalists, from Malinowski to Durkheim. For those, such as Parsons, who also stressed consensus, the twin assumptions led to a blurring of the issue of who is served by a contributing element.

Different individuals or groups, Merton contends, may receive different benefits from the same institution, custom, or practice. In fact, what is func-

---

[10]See Everett C. Hughes, "Dilemmas and Contradictions of Status," *American Journal of Sociology*, 50, 1944.

[11]Alvin W. Gouldner, "Cosmopolitans and Locals, I and II," *Administrative Science Quarterly*, 2 (1957-1958). Gouldner's critique of Parsons is further elaborated in Alvin W. Gouldner, *The Coming Crises of Western Sociology* (New York: Basic Books, 1970).

tional for some may be disfunctional for others. An increase in familial pride, for example, may be functionally adaptive for kinship groups, but may disrupt the solidarity of the total community. Proper application of a functional perspective, therefore, should analyze contributions in relation to specific units (such as families or status groups) rather than in relation to an amorphous society.[12]

Merton's critique actually has little meaning in regard to individualistic functionalism. Here the unit served by any social practice is clearly specified at the psychological or biological level. Further, and more importantly, the "basic needs" considered by such theories are presumed to be universal. Culture shapes their expression, but only minimal variability is typically assumed, especially within any specific society. Therefore, it isn't usually necessary to differentiate among subgroups of any kind because any functioning element is serving the same basic needs of everyone.

With respect to societal functionalism, however, Merton's critique illuminates an inherent problem. Characteristic of such theories is a nebulous view of the beneficiary of functional consequences. Thus, Durkheim wrote of benefits to the "organism" and the "social order." Contemporary expressions in this tradition write of benefits to the "social structure."[13] Such global benefits can only be posited along with either of the following assumptions: 1) Follow Durkheim's view that society is an entity separate from its parts. This assumption, as already discussed, raises teleological and tautological problems, which sometimes seem almost insurmountable, but may not necessarily make the approach heuristically fruitless; or, 2) Assume that all parts of a society share the same interests, and hence, are equally served by any given consequence. Modern societies are structurally too differentiated to assume that all parts share the same interests because they are alike. Therefore, Parsons's emphasis upon value consensus, despite structural heterogeneity, has been the more viable alternative for theories that assume all parts of a society are equally served by functioning elements.

In conclusion, the criticism that functionalism has overplayed consensus and integration while underplaying conflict seems well taken. One need not be a conflict theorist to recognize that strain, tension, and conflict are pervasive and endemic social processes. As Van Den Berghe states: "While societies do indeed show a tendency toward stability, equilibrium and consensus, they simultaneously generate within themselves the opposites of these."[14] Even

[12]Merton, *Social Theory and Social Structure*.

[13]See, for example, Arthur L. Stinchcombe, *Constructing Social Theories* (New York: Harcourt Brace, 1968).

[14]Pierre L. Van Den Berghe, "Dialectic and Functionalism," *American Sociological Review*, 28 (1968), p. 696. An approach that provides a more balanced treatment of conflict and consensus is provided by Louis Kriesberg, *The Sociology of Social Conflicts* (Englewood Cliffs, N.J.: Prentice-Hall, 1973).

Parsons recognized the simultaneous existence of such opposite forces when he stated that, "a complex social system is not either stabilized or changing"; it is "always both."[15] However, Parsons did not offer this statement until the eleventh, and final, chapter of his book on social systems. The first 500 pages of the book present a picture of society that stresses stability and integration. They were emphasized, he claims, to provide a model of stability and consensus from which change and disorganization could be analyzed. Without such a fixed referent, how could change be inferred? It was too little, too late, though. The model of society constructed in the first 500 pages was one in which conflict and change could not easily be viewed as fundamental social processes.

For the most part, our discussion of Parsons has been based upon his writings of the 1940s and 1950s; mid-century was the time of his greatest influence. The emphasis in Parsons's most recent books, however, has undergone some modifications. The analysis of more complex, modern societies has been stressed, leading to correspondingly greater attention to processes of change and conflict.

Specifically, change has been regarded as an evolutionary process primarily involving increased social differentiation. With increased differentiation, change is now seen to involve the possibility of structurally based conflicts that result, for example, from the attempts of various groups to protect vested class interests. Hence, value consensus and social integration are viewed as more problematic than in earlier writings. With increased differentiation, Parsons states, system loyalties cannot be assumed, which presents "a major problem of integration."[16]

Despite a more continuous and explicit interest in change and conflict, Parsons's recent writing retains many of its former emphases. The primary function of the social system, for example, is still considered to be integration; and the social system is still regarded as linked to the cultural system via institutionalization, and to the personal system via internalization. Furthermore, value consensus continues to be stressed and viewed as the foundation of legitimacy, even if such consensus is considered to be somewhat more problematic.

Parsons's earlier writing provided a point of departure for many critics who claimed that functionalists generally were unable to treat change or conflict in their conceptual scheme. In this regard the critics were equally guilty of exaggeration and overgeneralization. Van Den Berghe, for example, claims it was Radcliffe-Brown who "more than anyone else . . . blinded functionalism to the conflicts and contradictions inherent in the social

---

[15]Talcott Parsons, *The Social System* (New York: Free Press, 1950), p. 503.

[16]Talcott Parsons, *The System of Modern Societies* (Englewood Cliffs, N.J.: Prentice-Hall, 1971), p. 12.

structure.''[17] He viewed patterned relationships, such as joking and avoidance, as providing the answer to Hobbes's question; they made a stable society possible. While he did not view these patterned relationships as totally eradicating strain and conflict from social life, he also did not pay a great deal of attention to situations in which they did not function successfully, nor to the conditions that generated conflicts of differing intensity and form. Thus, the qualified criticism remains valid; namely, that functional theories have tended to tilt conceptually toward an exaggerated emphasis upon consensus and other related characteristics, such as integration and stability.

Choosing Sides

All functional theories are oriented to the analysis of how elements contribute to the perpetuation of a system. Elements of a potentially revolutionary nature, therefore, are likely to be implicitly evaluated as threats to the system, or else overlooked entirely, in functional analyses. After all, they do not provide answers to Hobbes's question. Critics have charged, as a result, that functionalism is politically and ideologically conservative; that it tends to support "establishments," regardless of their specific nature.[18] Thus, the functionalist sociologist, while pretending to be detached or neutral, is in effect taking sides.

In his famous critique of functionalism, Merton argues that the perspective is not *intrinsically* conservative. While acknowledging the above criticisms, he contends that functionalism is also intrinsically radical.[19] His best argument is that a functional analysis, by emphasizing the consequences of institutions, strips away from these institutions any inherent moral value. A family system that is highly valued in a society, for example, is not approached by a functionalist as being inherently good. It is critically evaluated according to its current consequences. Thus, Merton concludes, functional analyses can exert radical influences by theoretical attacks upon traditional institutions. Merton has also illustrated the obverse; namely, how widely disapproved of practices, such as ward politics, can contribute to social stability. Here again the implications of functional analyses can be radical in that they run counter to prevailing views.

[17]Van Den Berghe, "Dialectic and Functionalism," p. 696. For an example of criticism that totally rejects functionalism's ability to analyze phenomenon such as war and conflict, see Joan Smith, "The Failure of Functionalism," *Philosophy of Social Science,* 5 (1975).

[18]For statements of functionalism's conservative bias, see Bernard Barber, "Structural-Functional Analysis," *American Sociological Review,* 21 (1956); and Don Martindale, *The Nature and Types of Sociological Theory* (Boston: Houghton Mifflin, 1960).

[19]Merton, *Social Theory and Social Structure,* especially pp. 40-41.

Merton's points are well taken, but they overlook the fact that functional theories have rarely provided support for radical movements, while they have frequently supported establishments. The bias has been there, even if it is not intrinsic to the approach. There are two major reasons for this historic bias: first, the hesitancy of accepting survivals leads to efforts to find current benefits associated with every institution; and second, a classical conservative fear of change because of the difficulty in forecasting its eventual repercussions in a functionally integrated system. Each of these reasons will be examined in turn.

The hesitancy to accept survivals means that functionally-oriented sociologists approach concrete institutions expecting to find functional consequences. Their view of society from a "wisdom of the body" metaphor leads them to expect that the institution would not have persisted if it did not contribute something. As a representative illustration, consider the functional analysis of the Coronation in Britain by Shils and Young. "There is a recurrent need in men," they begin, "to reaffirm the rightness of the moral rules by which they live. . . ."[20] Note this assumption of moral consensus in a society as the authors' point of departure. (Shils was, it will be recalled, an early collaborator of Parsons's.) However, consensus is also viewed as problematic, and the people's recognition of the Queen is seen as reinforcing that consensus. The Queen's oath, the Archbishop's presentation of the Holy Bible, and other aspects of the Coronation are all seen by Shils and Young as reaffirmations of collective values. Each step relegates the Queen's private morality to a place beneath that of public standards. The Coronation, they conclude, provides the entire society with an "intensive contact" with collective morality. Thus, it is "a great act of national communion."[21]

All the specific inferences and assumptions made by Shils and Young are open to further questioning; but they do not provide the critical issue here. What should be noted is that this approach is supportive of the Coronation. It might easily be interpreted as a plea for the continuation of the Coronation. However, consider how different a sociological analysis might be if it started out by rejecting the assumption of moral consensus in Britain! Such an analysis would recognize that many persons in that country were opposed to the Royal Family and all its attendant trappings, such as the Coronation. After all, the millions of pounds allocated annually for the support of this institution might be better spent; and the royalty can be viewed as vestiges of a stratification system in which the privileges of ascribed ranks were extensive. From initial assumptions such as these, the Coronation might be viewed as serving the interests of some segments of British society at the expense of others. The

[20]Edward Shils and Michael Young, "The Meaning of the Coronation," *Sociological Review*, 1 (1953), p. 67.
[21]*Ibid.*, p. 81.

functional analysis of Shils and Young, from this non-consensus perspective, has a conservative stance. It is pro-Coronation, and we can therefore deduce that it is supportive of those segments of the society that are served by the continuation of the institution.[22]

It must also be noted, however, that functional analyses are not necessarily doomed to this conservative ideological bias. Such conservativism could be avoided by adherence to Merton's point that functional explanations should specify the unit that is served. A functional analysis that was sensitive to different units could actually be a powerful intellectual tool for showing the consequences of practices, such as the Coronation, for groups in the society with different vested interests. This type of analysis would be free of a conservative underpinning and in its infrastructure would actually resemble a Marxian approach.[23] Functionalism has not often been applied in this manner, though, and for societal functionalism this differentiation of units presents some inherent problems. Perhaps for this reason, "radical critiques" of functionalism have treated Malinowski's individualistic functionalism more kindly. His version of functionalism has been viewed as more congruent with a Marxian approach because it permits examination of the way inherent contradictions in a society "frustrate" man's basic nature.[24]

A second reason for the typically conservative bias of functionalism, as noted earlier, has been the tendency of its proponents to share the classically conservative fear of change. Assuming society to be functionally integrated leads to an emphasis upon the continuing repercussions of even seemingly minor changes. Just the innovation of stone axes, for example, may ultimately sound a society's death bell. Thus, tracing the repercussions of change was seen by Parsons, for example, as the crucial question in his earlier theory of social change, and the existence of manifold reverberations was intimately tied to the very conception of society as a social system.[25]

Gouldner's analysis of Malinowski's theoretical contribution to British colonial policies is an example of how uneasiness toward change has made functional theories more susceptible to a conservative bias. Malinowski's romantic view of native practices, combined with a warning that any change would have unpredictable consequences, was highly compatible with the colonial goal of maintaining European dominance. It meant that native societies would be protected much like endangered species, and this clearly meant no political autonomy and no "modernization." While some functionalists,

---

[22]For further discussion of the political aspects involved in the definition of functions, see Howard Becker, *Outsiders* (New York: Free Press, 1963).

[23]For a comparison of the ideological orientations of functionalism and dialectical materialism, see Merton, *Social Theory and Social Structure*, pp. 41-43.

[24]See, for example, Al Szymanski, "Malinowski, Marx and Functionalism," *The Insurgent Sociologist*, 2 (1972).

[25]Parsons, *The System of Modern Societies*, p. 494.

Gouldner concludes, "conceived it as their societal task to educate colonial administrators, none thought it their duty to tutor native revolutionaries.[26]

Fear of change is also related to the functionalists' tendency to construct lists of functional imperatives. As discussed in the preceding chapter, these lists typically included such necessities as the legitimation of authority, social control, and the socialization of youngsters into traditional roles. In considering change, one of the first questions Parsons suggests is: will the change violate any of the imperatives? For theorists who think in terms of prerequisites, a cautious and timid approach to change seems almost inevitable. However, the influence of the functional imperative argument upon conservative thinking is not strong, mostly because few people have taken these lists seriously.[27] All of the worst features of functional theories—teleology, tautology, conceptual ambiguity, etc.—are most pronounced in the case of the functional imperative viewpoints.[28]

In conclusion, there is substantial basis to the charge that functional theories tend to be conservative in their implications. Perhaps their conservatism is not intrinsic, but questioning the contribution made by elements for the maintenance of a system is not likely to generate much appreciation for radical or revolutionary forces in a society. However, it must also be recognized that there are strong and pervasive conservative forces in most societies, and functionalism is well-suited to their analysis. In other words, while critics within sociology are attacking functionalism's conservativeness, I think it is often society's conservativeness that they object to, and functionalism in this case is merely an accurate conceptual picture of the society. It seems clear, though, that functional sociologists have provided intellectual justification for the conservative elements, and that they have chosen sides, although often inadvertently.

Throughout the middle decades of this century, when functionalism —Parsonian functionalism, in particular—had overriding dominance in sociological theory, it gave a conservative tilt to the entire discipline. This seems less true today when a variety of perspectives abound, although many would still argue that functionalism is so pervasive that without it nothing would be left of sociological theory.

[26]Gouldner, *The Coming Crises of Western Sociology* (New York: Basic Books, 1970), p. 132. Gouldner's argument is not that sociology should be on either side. Rather, sociologists must be aware of these political influences in order to "protect" sociology. Alvin W. Gouldner, *For Sociology: Renewal and Critique in Sociology Today* (New York: Basic Books, 1973).

[27]Goode states: "Although this type of listing has frequently annoyed anti-functionalists, it has played a very minor role . . . "William J. Goode, *Explorations in Social Theory* (New York: Oxford University Press, 1973), p. 84.

[28]For a critical analysis of the functional imperative argument, see George C. Homans, "Structural, Functional and Psychological Theories," in *Systems, Change and Conflict*, eds. Nicholas J. Demerath III and Richard A. Peterson (New York: Free Press, 1967).

Is That All There Is?

One important reason for the few explicit defenses of functionalism over the years has been a tendency for the attackers to attribute to functionalism a set of unqualified assumptions that no one had really made in the first place. Thus, in metaphoric terms, functionalism is like a besieged palace with nobody inside the castle defending it.[29] From within this framework the criticism has a mythical quality; if there is a palace, that palace is the discipline of sociology rather than functionalism. However, there is no distinction between the two because functional and sociological analyses are seen as synonymous.

Their equivalence, in recent decades, has been openly debated. For example, it has been a favored topic in presidential addresses to the American Sociological Association. Kingsley Davis's address in 1959 was the most influential statement made by a person in this position. To demonstrate their virtually total overlap, Davis conceptualized the residual category of "non-functionalism" and proceeded to argue that whatever types of analyses were non-functional were also non-sociological; that is, if the discipline was properly defined.

Davis closely followed Durkheim's societal functionalism position and saw the residual category as containing two types of analyses:

1) Reductionistic views, which attempt to explain social patterns by inferring psychological or biological factors. Following Durkheim, Davis contended that reductionistic analyses were non-functional and non-sociological as well.

2) Radical or abstracted empiricism, which presents sheer description or statistics only. To be functional, an analysis must explain concrete phenomena in reference to abstract principles. It must, in other words, be comparative analysis to be either functional or properly sociological.

Finally, Davis concluded, in every discipline scientists do comparative analyses of variables that are theoretically relevant to their disciplines. Functionalists in sociology, therefore, are not a special breed; those who think otherwise believe in a myth.[30]

Throughout the turbulent 1960s and into the 1970s, however, other sociologists contended that functionalism led to choosing sides, to a preoccupation with consensus and stability, etc. And in their critical postures they did not see themselves as Don Quixotes. We have already considered most of these criticisms; therefore, let us focus specifically upon the two types of

[29]William J. Goode, *Explorations in Social Theory.* See especially chapter three on functionalism, "the empty castle."

[30]Kingsley Davis, "The Myth of Functional Analyses as a Special Method in Sociology and Anthropology," *American Sociological Review,* 24 (1959).

"residual sociology" identified by Davis: reductionistic and sheer description.

George Homans's presidential address in 1964 focused upon the reductionism issue, but also raised the larger question of whether Davis viewed functionalism as an empirical method or as a theory. Homans stresses this distinction, viewing the method as emphasizing structural interrelations and consequences (rather than causes). He claims to be a methodological functionalist himself and says that, perhaps, all sociologists are methodological functionalists, as Davis contended. But functional theorists? That is another matter, and Homans, for one, asks to be "counted out."

His major objection is to the societal functionalist tradition, which treats roles rather than people as constituting the most elementary unit of a social system. Hence, the title of Homans's paper: "Bringing Men Back In."[31] A theory's reason for being, he states, is to explain something; and it accomplishes this through a series of propositions. Functionalism fails in its charter as a theory, Homans asserts, because it is unable to explain anything. And it does not qualify as a theory in form either, because it lacks systematic propositions. Both deficiencies occur, he concludes, because the functionalists begin with roles rather than people. If, in a concrete analysis, they are ever able to explain anything, it is because they inadvertently "bring men back in" to their studies.

In order to illustrate this contention, Homans offers a restatement of Smelser's study of changes in the British textile industry.[32] Smelser introduced the study in functional terms, viewing the textile industry as part of a larger social system that was attempting to maintain some type of equilibrium in the face of rapid technological changes. However, Smelser's ultimate explanations for the changes, according to Homans, focus upon people's perceptions, motivations, strivings for financial gain, etc. These psychological characteristics of human beings actually account for why certain innovations occurred, according to Homans's view; and the functional imagery employed by Smelser, dealing with the characteristics of societies and groups, is just "so much window-dressing." Thus, "the only general propositions of sociology are, in fact, psychological."[33]

Homans's critique, in my estimation, suffers from a number of shortcomings. To begin with, it is internally inconsistent. He concedes that showing *structural* interrelations is an important aspect of both functional—and, more generally, sociological—analysis. It is precisely for this reason that sociologists (including functionalists) have utilized roles rather than individuals as the primary units of analysis. How Homans's reductionistic emphasis

[31] George Homans, "Bringing Men Back In," *American Sociological Review,* 29 (1964).

[32] See Neil Smelser, *Social Change in the Industrial Revolution* (Chicago: University of Chicago, 1959).

[33] Homans, "Bringing Men Back In," p. 817.

will transcend this unit of analysis problem and address questions at an institutional level is not clear. In addition, this insistent reductionism disregards the ways in which a social structure influences or determines people's perceptions, and the like. This is already familiar ground to us, however; see the earlier discussion of Radcliffe-Brown's reaction to Malinowski in Chapter 2.

Finally, Homans's position on the form of theory takes the viewpoint we have previously identified as "reconstructed logic." His position overlooks the fact that Smelser was able to offer good explanations—in Homans's opinion—and that he utilized a functional perspective to derive his conclusions. Therefore, didn't functionalism perform adequately as a "logic in use"?

It must be noted here that functionalism was not the only way to reach such conclusions; that is, it did not provide the only perspective by which explanations could be deduced. Homans, for example, offers a series of "people" propositions that could have led an investigator to highly similar explanations. We must, therefore, disagree with Davis's contention that functionalism is "all there is."

Davis's contention is also weakened by changes that have been occurring in other "schools of thought" within sociology since his statement in 1959. His second residual category, it will be recalled, contained historical and ethnographic accounts that involved sheer description without any macro-comparative analysis. Labeling theory, for example, at the time of Davis's remarks, was rather exclusively directed at descriptions of how usually deviant social identities become affixed to individuals. From this base, however, at least some labeling theorists have gone on to ask comparative questions about groups.[34] Thus, functionalism does not have a lock on this type of analysis. Other approaches can be shown to share some features in common with functionalism, such as a concern with consequences (of labels, for example). These resemblances, however, are too superficial to support the assertion that functionalism is "all there is." Furthermore, many of these other schools—partly because they have been developed in opposition to functionalism—involve a less conservative stance toward the social order, thereby making sociological theory, in total, more eclectic.

Conclusion

Any science at any point in its history, according to Kuhn, tends to be dominated by a paradigm: a conception of what constitutes the appropriate subject matter of the discipline, how it ought to be studied, and what an

[34]See, for example, Paul G. Schervish, "The Labeling Perspective," *American Sociologist,* 8 (1973).

investigator is likely to find.[35] Thus, it is a very general image or perspective in which theories and methods are subsumed. If we view functionalism as such a paradigm, and I think that it is, a number of ambiguities can, at least, be better understood. An example of one ambiguity that can be clarified is raised by Homans and other theorists: is functionalism a theory or a method? From this perspective, functionalism is neither. It is a paradigm, which includes both theory and method.

Scientific "revolutions" occur, Kuhn continues, when one paradigm replaces another in the preeminent position within a discipline. There is little question but that functionalism was the dominant paradigm in the middle decades of the twentieth century. Contemporary sociology, by contrast, has been described as a multiple paradigm science, though there is less than complete agreement concerning which perspectives constitute the major paradigms.[36] Almost regardless of what also might be included, however, functionalism is considered one of these dominant paradigms.

The jockeying for positions of preeminence among paradigms is viewed by Kuhn as partially political and partially scientific.[37] It is political in that which paradigm wins may be dependent upon the relative power of its supporters within the discipline. It is scientific in that the winning paradigm may be the most fruitful or accurate in its predictions. However, popular conceptions aside, scientific revolutions are viewed as generated by more than simple questions of scientific merit.

With respect to functionalism's fall from sole occupant of the pedestal, much of the attendant debate was more political in an ideological sense than is usually the case in a scientific revolution. The perceived tendency for functionalist sociologists to side with establishments, for example, became an important rallying point for the development of counter perspectives by ideologically more liberal sociologists.[38] As Goode has noted, "the bitterest attacks [on functionalism] are mounted by politically dissident sociologists. . . ."[39] All of this is not meant to imply, however, that much of the criticism was not also well-founded, regardless of its ideological or political motivations. Thus, we have noted the problems of functionalism with respect to ambiguity, tautology, conceptual biases, and so on. Sociology moved from a functionally dominated to a multiple paradigm discipline for both political and scientific reasons.

[35]Thomas Kuhn, *The Structure of Scientific Revolutions* (Chicago: University of Chicago, 1962; second edition, 1970).

[36]George Ritzer, "Sociology: A Multiple Paradigm Science," *American Sociologist,* 10 (1975).

[37]Kuhn, *The Structure of Scientific Revolutions.* Kuhn places more emphasis upon the political in his first edition (1962), and more on the scientific in the later edition (1970).

[38]See, for example, Howard S. Becker, "Whose Side Are We On?" *Social Problems,* 14 (1967).

[39]Goode, *Explorations in Social Theory,* p. 65.

II
Substantive Applications of Functionalism

4

stratification

A functional theory of stratification was explicitly developed primarily during the decade between the late 1930s and 1940s. This functional perspective was shaped and defined mostly in the writings of Talcott Parsons, Kingsley Davis, Wilbert Moore, and to a lesser extent, W. Lloyd Warner. It is most often associated today with the names of Davis and Moore because their brief but succinct article of 1945 provided a very conspicuous statement of the theory. However, the origins of the theoretical approach can be traced backwards, notably to Durkheim, and there has been continuing debate regarding the perspective that has led to some refinements of the earlier statements.

In this chapter we will begin by briefly considering the existing relationships between this stratification theory and the earlier, more general functional perspectives. We will also review, in the final section of this chapter, a number of the recent attempts made to assess the theory empirically. We will also focus upon the writings of the theory's proponents and its critics.

The Functionalist Background

One clear connection between the functional theory of stratification and the earlier and more global functional theories was via the influence of Durkheim's writings on Talcott Parsons. The latter explicitly incorporated many aspects of Durkheim's view of society into his writings, and specifically, began his theories of stratification from Durkheim's position. For Durkheim, stratification involved a general ranking of individuals based upon "moral," or normative evaluations. In Parsons's view, this pervasive moral ranking became an important basis of others' expectations; that is, individuals were expected, by others, to behave in accordance with their location in the stratification system.[1]

The functional theory of stratification was also tied to the earlier and more global functional theories as a result of the personal influence of Radcliffe-Brown upon W. Lloyd Warner. The latter (along with Hollings-head) is generally recognized as the father of empirical studies of stratification. Warner's studies and those of his students, in "Yankee City," "Jonesville," and so one, established the pioneering standards of community stratification studies.

Warner's sociological studies of stratification in America were conducted after his anthropological research among the Murngin tribes of Australia.[2] This early research, conducted between 1926 and 1929, was directed by Radcliffe-Brown. His subsequent research on stratification involved an explicit attempt to view the then contemporary American community from the same functionalist theoretical perspective that Radcliffe-Brown, Malinowski, and others had previously developed in studying primitive societies.

These varied historical influences will become clearer as the nature of the functional theory of stratification is presented in the following pages. Let us turn now to Parsons, and examine his early writings on stratification.

Parsons's View

Parsons, as already noted, began with Durkheim's view of stratification as a moral evaluation. An individual's wealth, lineage, or some other salient characteristic was seen as utilized by others to place the individual into a ranked status. This status rank, in turn, was viewed as "structuring" others' expectations. Parsons's general tendency to assume consensus, as discussed

---

[1]Emile Durkheim, *Moral Education,* ed. by Everett Wilson (New York: Free Press, 1973). Parsons's utilization of Durkheim's perspective is clearly expressed in Talcott Parsons, *The Structure of Social Action* (New York: McGraw-Hill, 1937), especially chaps. 10-12.

[2]See W. Lloyd Warner, *A Black Civilization* (New York: Harper, 1937).

in Chapter 3, was also very evident in his assumption that everyone's evaluation of an individual's status would be highly congruent. If they were not, he argued, the stabilizing influence of shared normative expectations would be disrupted, producing, "a functionally impossible state of lack of integration of the social system."[3]

The primary criterion of an individual's status, he continued, is the individual's occupation.[4] Of greatest relevance to stratification is the wealth that is associated with an occupation; but Parsons also noted other potentially important occupationally-related attributes, such as prestige and influence. This occupationally-based stratification ranking was seen to be a male-dominated system, with a woman's status being derived from her husband's standing. Family solidarity, Parsons stated, requires that husband-wife competition in the status sector (i.e., the occupational realm) be minimized. Men, therefore, struggled for occupational attainments and women stayed home, working on their physical appearances.

The occupationally-based stratification system, as implied above, was also seen as resulting primarily from individual (i.e., male) achievements. Parsons recognized the possibility of high status—or any of the specific attributes that could result in high status—being based upon ascription; for example, inherited wealth or inherited social standing. In modern societies like the United States, however, such occurences were viewed as occuring only sporadically and as being of very minor significance. Thus, in Parsons's view, the stratification system was based upon consensual evaluations of male achievements in the occupational realm.

If evaluations of individuals are based primarily upon their occupations, then logically the next question is, What accounts for the consensual ranking of occupations? Parsons offered several answers to this question. For example, he identified the amounts of skill, required education, and authority over others as being potentially associated with the ranking of a position. Parsons's primary emphasis, however, was upon economic reward and value consensus. Thus, his major answer to the question of what accounts for an occupation's rank is: "the more highly valued jobs are the best paid."[5] As visualized in this way, wealth was a symbol of high ranking rather than a cause, an assertion later stressed by Davis and Moore.

Davis and Moore

The single most succinct presentation of the functional theory of stratification was offered by Davis and Moore in 1945. This article later came to be

---

[3]Talcott Parsons, "An Analytic Approach to the Theory of Stratification," *American Journal of Sociology*, 45 (1940), p. 843.

[4]*Ibid*. See also, Talcott Parsons, "The Professions and Social Structure," *Social Forces*, 17 (1939).

[5]Parsons, "An Analytic Approach," p. 857.

the "conspicuous statement" of the theory, and it partially accounts for their two names being used almost interchangeably with the theory itself. During the early 1940s they developed a rather lengthy manuscript to present the theory, and Davis alone published a brief article that outlined many of the major points.[6] Unfortunately, in light of the ensuing debate, they condensed the monograph into the brief, seven-page article in 1945.[7] Extensive criticism, accompanied by rejoinders and still more criticism, followed the presentation of the 1945 article, which was viewed not as concise, but as unqualified and dogmatic.

In the 1945 statement, Davis and Moore posed this central question: "why do different positions carry different degrees of prestige?" It is a different question, they insisted, to ask how specific individuals obtain those positions. Much of the ensuing criticism failed to appreciate the difference, however. Before considering this, though, let us first consider how the two theorists answered their basic question.

Differing degrees of prestige (and other types of rewards) are universal, they began, because every society must place the "proper" people into the most important positions and then motivate them to perform the duties associated with the positions. Unlike Parsons, they did not limit these inferences to achievement-oriented societies. All societies, competitive or not, were seen as having to absorb people into positions and then motivate them by differential rewards. Like Parsons, they saw this distribution of rewards as giving rise to a stratification system. Unlike Parsons, though, they did not equate functional importance with consensual values. Rather, following Durkheim, they viewed society as the sui generis, with stratification linked to its (i.e., the society's) needs. Thus, in what was to become perhaps the most problematic phrase in the essay, Davis and Moore asserted that stratification was an "unconsciously evolved" mechanism through which *societies* went about assuring that the best qualified people wound up in the most important jobs.

While continuing to stress shared values, Parsons's later writings on stratification also emphasized the "actual" functional importance of positions. Like Davis's and Moore's view, this defined society as a structural "reality" with inherent, endemic needs.

In achievement-oriented societies, Parsons argued, there must be a strong relationship between the division of labor (which he termed, "the instrumental structure") and the distribution of rewards. Those roles that involve the greatest degree of "responsibility for the affairs of the collectivity" will have more "facilities" at their disposal, and these facilities are "in

[6]Kingsley Davis, "A Conceptual Analysis of Stratification," *American Sociological Review,* 7 (1942).

[7]Kingsley Davis and Wilbert E. Moore, "Some Principles of Stratification," *American Sociological Review,* 10 (1945). A later elaboration was provided in Kingsley Davis, *Human Society* (New York: Macmillan, 1948).

themselves rewards."[8] In essence, if a division of labor and achievement values exists within a society, there must be differential rewards for positions of varying competence and responsibility. These differences in reward are the basis of the stratification system. "The only way to avoid this," Parsons concludes, would be to deny "differences of competence or responsibility, including denial of their functional relevance."[9]

One of the fundamental arguments made by Parsons is that stratification is necessary, and also desirable, for a complex, achievement-oriented society: necessary because it allocates rewards and "connections" to positions according to the amount of collective responsibility entrusted to them; and desirable because this arrangement permits the entire system to function effectively. Thus, Parsons noted that industrial managers in Russia were part of the "intelligentsia" and received more of all types of rewards than ordinary workers. Their greater rewards are not valued as legitimate within the Soviet system, and the same inequities in the United States would be considered "capitalistic." Nevertheless, unequal rewards accrue to managers in both societies, and "a sociologist is at least entitled to be skeptical" that any ideology could change this pattern given "the essential structural situation" in a complex society.[10]

During this same time period, a highly similar view of the function of stratification provided one of Warner's basic assumptions. He suggested that there is a great need for coordination in heterogenous and differentiated (i.e., modern) societies. Those positions that are most entrusted with responsibilities for coordinating and directing activities in the society will be the best rewarded and most prestigeful positions.[11]

In addition to functional importance, Davis and Moore also proposed that the ranking and rewards of a position were determined by the relative scarcity of qualified personnel. If the obligations of a position require substantial amounts of innate talent or extensive training, then greater rewards will have to be associated with the position in order to induce a scarce pool of potential incumbents to seek the position. What they did not explicitly consider was how functional importance and scarcity *jointly* contributed to reward. On the one hand they imply independent effects; that is, either great functional importance or high scarcity is sufficient to ensure high rewards. On the other hand, they imply a multiplicative interaction between the two variables; that is, if either one is very low it can mitigate the effects of the other. Their cited example, modern physicians, does not help to clarify the situation

[8]Talcott Parsons, *The Social System* (New York: Free Press, 1951), p. 159.

[9]*Ibid.*, p. 159.

[10]*Ibid.*, p. 160.

[11]W. Lloyd Warner, *Social Classes in America* (Chicago: Science Research Associates, 1949). See also, W. Lloyd Warner and P. S. Lunt, *The Status System of a Modern Community* (New Haven: Yale University Press, 1947).

because they view the position of medical doctor as functionally important, and the training as so rigorous that few can qualify. Thus, both functional importance and relative scarcity of qualified personnel have congruent consequences for the rewards of physicians. It is unclear what Davis and Moore think would happen if the effects of both variables were not congruent.

Much to the credit of the theory, they did explicitly recognize that although stratification was universal, its specific form would vary in relation to "major societal functions." Here Parsons's influence is most apparent as Davis and Moore begin by considering the universality of religion and how increasing secularization is related to changes in the ranking of religious practitioners. As discussed in Chapter 3, Parsons's theory emphasizes the role of religion, and moral values in general, in the integration of society. A declining cultural emphasis on religion was seen as occurring in response to the "rationalizing" influence of "progressive forces." Thus, the initial concern of Davis and Moore with the effects of religion on stratification systems shows a clear Parsonian influence.

They proposed that in medieval types of societies, specifically, the organized priesthood is very high in prestige. This high rank is due to the functional importance of religion in societies of this type, where an "unlettered" population is highly "credulous." Given the importance of religious ritual in such "sacred" societies, it may be surprising, they note, that the position of the priest is not ranked even more highly than it is. What tempers their status, Davis and Moore note, is the ease with which anyone can claim to be in communication with deities, without fear of rebuke. Thus, there is a limited pool of eligibles in such societies only if literacy is a prerequisite. Therefore, the highest ranking of priests occurs when the priestly guild itself rigidly controls access to the profession.

Similarly, Davis and Moore go on to describe other possible variations in ranking due to changes in government, relations to the means of production, and technical knowledge. They conclude that actual stratification systems can be a number of polar types, varying in equalitarianism, opportunities for mobility, degrees of stratum solidarity, and so on. However, even though the form may vary, functional importance and relative scarcity are seen as the basic principles of stratification.

The Criticisms

Since its initial presentation, the functional theory (Davis and Moore; in particular) has been the target of numerous criticisms. Some have been well-founded and have led to intellectual exchanges in which theoretical issues have been clarified. Some of the criticism, however, has led only to digressions because it has been based upon misunderstandings generated by the complexity of the theory and its overly brief presentation. The most visible and continuing critic is Melvin Tumin, who has authored about a dozen papers on varied aspects of the Davis-Moore theory.[12]

measuring functionality

One of the first issues raised by Tumin involves "the calculus of functionality." That is, do non-subjective criteria exist to enable the functional importance of a position to be assessed? He argues that Davis and Moore utilize non-demonstrable values in claiming, for example, that engineers are functionally more important to a factory than are unskilled workers. In the long run, Tumin asks, must not everyone in an enterprise be adequately motivated? Davis agrees with this statement, but points out that engineers must receive more training, which would not occur unless their work was more important and, hence, better rewarded. However, the ambiguity raised by the notion of functional importance continued to generate debate. Simpson, for example, has argued that the prestige of garbage collectors is lower than what should be accorded to them in light of the serious problem that uncollected refuse would present to a society.[13] Furthermore, the noxious aspects of garbage collecting also may serve to limit the potential supply of recruits.

In considering Simpson's criticism from the perspective of Davis and Moore, Huaco points out that the functional theory proposed only a tendency for rewards to vary with functional importance; its correspondence with actual rewards need not be perfect.[14] In other words, garbage collectors may be an exception, and few theories can avoid any exceptions. This reply does not resolve the problem, though. As long as functional importance is not operationally defined, people are free to disagree with each other's intuitive evaluations. Thus, there will be no basis for arguing about whether the number of exceptions is greater or smaller than the number of non-exceptions.

Davis and More offered two guides: the uniqueness of a position, and the number of other positions that are dependent on it. Consider a baseball team in this regard. The pitcher's position is more unique than a right fielder's, for example, and this could be demonstrated by showing greater interchangeability of positions. Specifically, more people who play other positions (such as first base or left field) may also play right field as compared to those who play other positions and also pitch. The dependence of other positions, including the dependence of the entire team, is also greater on the pitcher than on the right fielder. Based on these indications of functional

[12]Several of these papers are reproduced in Melvin M. Tumin, ed., *Readings on Social Stratification* (Englewood Cliffs, N.J.: Prentice-Hall, 1970). For a listing of these papers, see p. 405.

[13]Richard L. Simpson, "A Modification of the Functional Theory of Stratification." *Social Forces,* 35 (1956).

[14]George A. Huaco, "The Functionalist Theory of Stratification," *Inquiry,* 9 (1966).

[15]These and other illustrations are analyzed in Mark Abrahamson, "Talent Complementarity and Organizational Stratification," *Administrative Science Quarterly,* 18 (1973).

importance and relative uniqueness, the material and symbolic rewards of pitchers should exceed those of right fielders.[15]

Pursuing the same pattern of reasoning, the functional importance of quarterbacks could be differentiated from that of offensive linemen; soloists could be differentiated from members of symphony orchestras; and so on. However, note that in each case positions are ranked in functional importance relative to other positions within the same organization. Inter-organizational comparisons also are suggested by the Davis-Moore theory; that is, offensive linesmen in football teams should be compared to musicians in orchestras. This latter comparison requires a basis for evaluating the relative functional importance of different components of a society. It is the type of analysis Davis and Moore themselves attempted to develop in examining the changing role of priests as the religious institution within societies changed.

Questions concerning exactly how the concept of functional importance can and should be measured continue to be baffling. Some sociologists, following the Parsonian viewpoint, have regarded agreed-upon evaluations as indicators of functional importance. However, people may not be accurate in their perceptions of a position's actual contributions to society, and it is clearly the latter that is emphasized, especially by Davis and Moore. Other sociologists have attempted to logically deduce circumstances under which functional importance would vary, and try in this way to infer functional importance directly; that is, without intervening perceptions. However, the inferences that they have viewed as logical have not been universally regarded as valid. In a later section of this chapter we shall review empirical studies of the theory, and return for a more detailed consideration of the measurement of functional importance.

is stratification functional?

A second major criticism, from Tumin and others, is directed at the presumed inevitability and desirability of stratification systems. The more stratified a society is, he claims, the more likely are talented, lower-standing persons to go "undiscovered." Because their access to mobility channels often is denied, they are not likely to develop their talents. From the standpoint of the society, he concludes, this is hardly functional.[16] Furthermore, Tumin questions whether rewards must be viewed as the best way or the only way of recruiting appropriate talent: what about intrinsic work satisfaction as an alternative?

[16]Tumin has been joined by a number of others who contend that stratification has similar kinds of dysfunctional consequences. See, for example, Dennis H. Wrong, "The Functionalist Theory of Stratification," *American Sociological Review*, 24 (1959); Walter Buckley, "On Equitable Inequality," *American Sociological Review*, 28 (1963); and Randall Collins, "A Conflict Theory of Sexual Stratification," *Social Problems*, 19 (1971).

These motivational assumptions are at the heart of the criticism offered by the Polish sociologist, Wlodzimierz Wesolowski. The view held by Davis and Moore concerning the indispensability of rewards is based, Wesolowski contends, on the assumption that human nature is characterized both by selfish, materialistic drives, and by laziness.[17] Rewards are correspondingly viewed as the necessary energizing mechanisms. Davis and Moore have disregarded, he concludes, the impact of cultural values on motives of behavior. This leads them to be insensitive to the possibilities, in some cultures, of training people to fill important positions without their reckoning on future material advantages.

The most important issue raised by this line of criticism has been termed, "the strangulation of talent." For Davis and Moore and other functionalists, stratification was viewed as having positive consequences, and society was identified as the beneficiary. The "strangulation of talent" argument contends that stratification is dysfunctional because it leads to the underdevelopment and underutilization of potential ability. This strangulation occurs particularly among groups in a society that lack access to high-ranking positions (and intervening institutions that offer mobility) by virtue of ascribed characteristics: birth into low-standing (and hence, low-resource) families; minority group status; etc. The critics contend that in the long-run, any institutional structure (such as stratification) that limits the pool of eligibles must be dysfunctional to the society.

This strangulation of talent argument has elicited two kinds of replies from Davis and Moore: one a concession, the other a clarification. They have conceded that their theory might be partially limited in applicability to achievement-oriented societies.[18] That is, to the degree that ascribed characteristics determine positions, the principles of stratification they posited may be of lesser importance. This partial limitation, it should be noted, brought the Davis and Moore position into still closer accord with the positions of Parsons and Warner. However, along with the concession came an important clarification. The Davis-Moore theory was primarily an attempt, the critics were reminded, to account for the ranking of *positions*. It did not purport to explain how individuals or groups attained such positions; and it remains a separate issue.[19] (Parsons, it will be recalled, did treat these two issues simultaneously; but the critics were specifically addressing the Davis-Moore version.) Thus, even in a highly ascribed caste system, the differential ranking of specific castes can be explained by the theory, even if individuals are sorted into castes

[17] Wlodzimierz Wesolowski, "Some Notes on the Functional Theory of Stratification," *Polish Sociological Bulletin,* 3-4 (1962).

[18] Davis initially modified their joint position in *Human Society*. In reply to continuing criticism, this limitation was explicitly repeated. See Kingsley Davis, "Reply," *American Sociological Review,* 18 (1953); and Wilbert E. Moore, "But Some Are More Equal Than Others," *American Sociological Review,* 28 (1963).

[19] See especially Davis, "Reply."

ascriptively. Even in a caste system, though, there is some vertical mobility and here the stratification system is likely to play the part attributed to it; namely, the drawing of the most capable people into the most important positions.

Closely related to the "strangulation of talent" issue are the questions of motivation and inevitability of stratification. In part, positions on these questions are ideological and based, as Wesolowski noted, on fundamental conceptions of the nature of the human being and society. From this perspective, the functional theory of stratification can be viewed as taking a conservative stance, emphasizing the necessity of social institutions.[20] The critics, from this vantage point, may be seen as arguing from a more liberal position that social institutions corrupt people's "inherent goodness."

Clearly, people will be either attracted to or repelled by the functional theory according to their general ideological predispositions. This is an element of the theory that is not directly subject to logical or empirical tests. This does not mean, however, that assessment of the theory can never involve more than ideological discourse. The functional theory also implies certain empirical consequences, Stinchcombe has noted, and their truth or falsity is subject to empirical assessments as well as the ideological controversy.[21]

Empirical Assessments

In order to illustrate this contention, Stinchcombe formulated a number of hypotheses that he derived from the Davis-Moore article. No relevant data were presented; rather, they were presented as *testable* hypotheses, to provide illustrations of how the theory's empirical utilities might be assessed.

He proposed, for example, that in Western European democracies, kings have declined in their political importance relative to the importance of parliaments. Correspondingly, he argued, the rewards of kings have declined; they have less power and less wealth, for example. Given their reduced importance, the role requirements to be king also have diminished. This is indicated, for Stinchcombe, by more ascriptive successions to the kingships, fewer palace revolutions to abolish incompetent kings, and fewer contested successions to thrones.

The changes that Stinchcombe hypothesizes are amenable to historical study. The wealth of kings, the number of contested successions, and the like, all could be examined historically. The trends could be shown to be either consistent or inconsistent with the hypothesis. However, note that the declining functional importance cannot be directly measured. Without such mea-

---

[20]Gerhard E. Lenski, *Power and Privilege* (New York: McGraw-Hill, 1966).

[21]Arthur L. Stinchcombe, "Some Empirical Consequences of the Davis-Moore Theory of Stratification," *American Sociological Review,* 28 (1963).

sures, it remains open to conjecture whether changes in the rewards or role requirements of kings are, in fact, due to changes in their functional importance.

Stinchcombe's clearest statement on this issue is that: "changes in the nature of role-requirements and of the rewards indicate a shift of functions."[22] Unfortunately, this is a tautology. Changes in rewards and requirements are theoretically assumed to be the *consequences* of changes in functional importance. They cannot, therefore, be utilized to indicate that changes in functional importance have occurred.

His other statement on this matter is based on the observation that kings may continue to have important ceremonial roles even though their political roles are more tangential. Their nonpolitical function probably leads them, "to be less important in the eyes of the people."[23] As we have previously noted, however, consensus is a very problematic indicator of functional importance. People's perceptions of importance, in other words, will not necessarily be closely related to the actual contribution of an individual position to a society. Thus, the familiar dilemma persists: how is functional importance to be measured?

One attempt to follow Stinchcombe's reasoning involved making a logical, albeit unproven, inference concerning fluctuations in functional importance. Specifically, it was assumed that times of peace and war altered the relative functional importance of military and civilian occupations. Positions directly involved in the war effort (e.g., army captain) should gain in relative rewards received during time of war, and suffer relative losses during peace times. The reverse was anticipated for civilian occupations (e.g., public school teacher).[24] Note, it was assumed that times of peace and war produced variations in the functional importance of positions; hence, peace and war times were correspondingly utilized as indicating differing degrees of functional importance. This type of measurement presumably reflected variations in actual functional importance, rather than people's perceptions of it; but the inference was supported only in this logical way, without direct evidence.

The reasoning behind the peace and war hypothesis should already be familiar to the reader. It is essentially the same reasoning as that followed by Davis and Moore in analyzing changes in the role of the priesthood, or Stinchcombe in explaining the decline of kingships. The data involved salary changes in matched occupations between 1939 and 1967. The matched occupations consisted of pairs of positions with equivalent prestige scores during the period in question; for example, army corporals and machine operators in factories were one such pair.

[22]*Ibid.*, p. 806. See also Arthur L. Stinchcombe, *Constructing Social Theories* (New York: Harcourt Brace, 1968), pp. 80-101.

[23]*Ibid.* See also Huaco, "The Functionalist Theory of Stratification."

[24]Mark Abrahamson, "Functionalism and the Functional Theory of Stratification," *American Journal of Sociology,* 78 (1973).

The average salary figures for the matched pairs covered three war-time periods (World War II, Korea, and Viet Nam) and the intervening times of peace. A comparison of salary increments indicated that military occupations typically gained, relative to civilian occupations, during times of war; the civilian occupations tended to experience the relative gain during times of peace. Thus, if it is assumed that the states of war and peace alter the functional importance of positions, then the results of this study are consistent with the theory's basic assertion that the magnitude of reward is a consequence of the functional importance of a position. (Relative scarcity, the other half of the Davis-Moore equation, was not explicitly considered in this study. However, expanded conscription during the times of war would increase the pool of eligibles, operating against the presumed effects of increased functional importance.)

Though his discussion of Stinchcombe substantially predated this study, Huaco anticipated the possibility of similar findings with skepticism. The coordination of positions during times of war, he proposes, creates "unusual" social system conditions; that is, situations that cannot be generalized. Huaco is apparently assuming that: 1) times of war are rare, statistically, which they are not, and/or 2) relationships that hold during times of war do not hold during times of peace. This latter objection may be true in some cases; however, it is questionable as a general theoretical statement, and it is not consistent with the findings of this study.

In addition, other critics have contended that this study actually indicates the "blindness" of the functional theory to questions of power, and they contend that the data could be reinterpreted so as to be inconsistent with the theory. A rejoinder denies the charges, of course, and presents further support for the theory.[25]

The best way to resolve questions of this type is by examining the results of other studies (i.e., replication). In this case, the best we can do is to examine the results of two other studies that have focused explicitly upon functional importance. In both cases it was found to be relatively inconsequential as a corollary of either prestige or reward. However, both of these studies utilized people's perceptions in measuring functional importance. Therefore, the degree to which findings will vary dramatically, solely as a result of different operational measures is not clear. Each of the studies also possesses other uncertainties, requiring that each be examined in some detail.

---

[25] See Marvin D. Leavy, "Comment on Abrahamson's 'Functionalism and the Functional Theory of Stratification'," and Beth Vanfossen and Robert Rhodes, "A Critique of Abrahamson's Assessment," *American Journal of Sociology,* 80 (1974). See also Mark Abrahamson, "In Defense of the Assessment," *American Journal of Sociology,* 80 (1974), pp. 732-38

perceptual studies

The first of the studies utilized a sample of only 185 United States high school students, who were asked to rank the functional importance, skill, prestige, and reward of 24 occupations.[26] This is a highly suspect sample, both in terms of composition and size; but let us review the findings. The investigators correlated the students' rankings of the four variables, and initially obtained results that were highly congruent with the Davis-Moore theory. Functional importance correlated with both prestige and reward (0.75 and 0.59, respectively); and the perceived skill required also correlated with both prestige and reward (0.92 and 0.89).

From the magnitude of these correlations it appeared that skill was more closely related to prestige and reward than was functional importance. Subsequent analysis of the data confirmed this quite strongly. Specifically, perceptions of functional importance and skill were themselves interrelated, so the investigators wished to examine the independent effects of each; that is, the effects of each when the other was held constant. Holding functional importance constant hardly influenced the other relationships. However, when skill was held constant, the correlations between functional importance and prestige, and functional importance and reward, were both very markedly lowered (to 0.37 and 0.19, respectively).

Although some of the variables were different (precluding precise comparability), highly similar results have been reported from a large national survey of over 1500 persons in Italy.[27] The respondents were asked the degree to which "the prestige of a person" was "augmented" by: wealth, occupational skill, the utility of a person's job, and family origin. The investigators reasoned that support would be provided for the Davis-Moore theory if most persons selected skill and utility (i.e., functional importance). They were, in fact, the most widely selected variables, and most respondents felt that they increased prestige by a considerable amount.

In another portion of the questionnaire, the investigators attemped to measure functional importance by asking respondents to rate the importance they believed others attributed to their own jobs. The other half of the Davis-Moore equation, scarcity, was indicated by the respondents' educational level. Then, these measures of functional importance and scarcity, or skill, were used to predict rewards. The latter was also a self-reported measure, involving income and respect. Their results indicated that both education

---

[26]Joseph Lopreato and Lionel S. Lewis, "An Analysis of Variables in the Functional Theory of Stratification," *Sociological Quarterly,* 4 (1963).

[27]Burke D. Grandjean and Frank D. Brown, "The Davis-Moore Theory and Perceptions of Stratification," *Social Forces,* 54 (1975).

and importance were independently related to rewards (combined $r^2 = .41$),[28] in line with theoretical expectations. However, the effects of the importance variable were not large, and they were substantially less than the effects of education.

This study is simply replete with measurement problems. Huge inferences are required to view the investigators' operationalizations as isomorphic with the concepts Davis and Moore articulated. Yet, the results are congruent with those obtained from the United States high school students, despite the sampling limitations of that study. Combined, the two studies suggest that Davis and Moore exaggerated the stratification consequences of functional importance, but not of scarcity or skill. However, this conclusion follows only if perceptions are accepted as the measure of functional importance. This problem of measurement, already a familiar one, seems to continually reappear at crucial times. Prior to attempting to offer some final conclusions concerning measurement, however, let us review two additional studies that followed somewhat different approaches.

critical tests

The studies that have been considered to this point have exclusively involved testing the aspects of the Davis-Moore formulation. They have not attempted to test the explanatory power of the functional theory in comparison with other competing perspectives. The latter type of assessment provides a much more critical test. The superiority of the latter type stems from the fact that different theories may share a variable or contain different variables that nevertheless correlate with each other. As a result, to focus solely upon one theory may lead to very short-sighted interpretations of research findings. In other words, results that are interpreted as supporting one theory might, if viewed from other perspectives, be seen equally to support another theory.

Under ideal circumstances, a critical test isolates a very specific problem for which the theories in question offer different predictions. (There may also be a number of situations where the theories offer congruent expectations.) Unfortunately, there have not been any well-executed critical tests involving the functional theory of stratification. The two efforts we will consider here have both made questionable inferences in confronting the familiar problems of measurement. And the data presented have been ambiguous in the sense that they could be interpreted in a variety of ways. They are the closest approximations to critical tests, however, and for that reason their results warrant consideration.

Both of these "almost-critical-tests" are alike in a number of respects. Both focused upon education and educational institutions and tried to infer

---

[28]Measured in this way, the results indicated an additive relationship between education and importance, with no interaction effects.

functional importance directly, rather than utilize the mediating variable of the perceptions of people. The studies were also alike in that they posited a similar alternative to the functional theory. Specifically, both proposed a market competition perspective as the alternative. From this point of view, the rewards of positions are seen as a consequence of groups' differential power in a competitive marketplace. Physicians, for example, would be viewed as highly rewarded because of their ability to limit access to medical schools and other "monopolistic" practices, rather than as a result of their functional importance to society.

The first of these studies examined the distribution of rewards, by faculty rank, in almost 100 doctorate-awarding universities. All of these schools, it may be assumed, are commited both to research and to teaching; but, their emphasis upon either is subject to variations. The national prestige of the university departments was found to be highly related to the total research productivity of the departments, and thus was utilized to place the schools into one of three categories according to their relative research versus teaching emphasis.[29]

Following Stinchcombe, teaching-oriented schools are examples of talent additive organizations in which the contributions of individual members to the final product are more nearly equal and alike. Symphony orchestras or assembly lines in factories are other examples. The research-oriented schools, by contrast, are examples of talent complementary organizations, which are characterized by marked differences in the importance of individual contributions. Athletic teams, or the performers in a motion picture, provide other examples. As a result of underlying differences in functional importance, Stinchcombe proposes that there will be greater differences in rewards within organizations where talent is complementary rather than additive.

Comparing faculty rank differences in average salaries in the research (i.e., complementary) and teaching (i.e., additive) schools produced the expected results. However, it was also recognized that the greater magnitude of inequality might be due to differences in market competition. Faculty in departments where there is great emphasis on research might be expected to publish more, establish more prominent reputations, and, hence, be more actively recruited by other universities. Thus, the greater salary differences in research-oriented departments might be due more to marketplace pressures than to functional importance.

In order to approximate a critical test of the two explanations, the nature of the salary discrepencies was examined further. Following the functional theory it was assumed that the skills and contributions of the most senior faculty, the full professors, would be of special importance. Therefore, the largest discrepancy in rewards should occur between those of full and associate professors. According to the market competition theory, the relative

---

[29]Abrahamson, "Talent Complementary and Organizational Stratification."

"marketability" of associate professors is emphasized. Unlike full professors, they can be offered a promotion by a university attempting to entice
them, and they tend to be younger and less reluctant to move. (In fact, looking
at the faculty rank at the school of origin, associate professors were found to
move about four times more often than full professors.) Assistant professors,
by contrast, are unproven commodities. Thus, the market competition theory
leads to the expectation that the ratio of associate to assistant professors'
salaries should exceed the ratio of full to associate professors' salaries.

The results strongly supported the functionalist perspective, as outlined
above. The full to associate professors' ratio was the greatest in all types of
universities, and especially in the highly research-oriented schools. (Both
perspectives predicted maximum differences in the high research schools.
They differed in where the largest differences were expected to occur.) What
weakens this study as a critical test, however, is the large number of inferences and assumptions that were made. To begin with, is the
complementary-additive distinction based upon the functional importance of
positions? Or is it based upon the skill they require?[30] Skill or talent has
consistently been shown to be related at least to perceived differences in
reward, while functional importance has not. Further, the specific differences
examined in the study of university salaries were not the ideal ones to be
isolated in a test of the two perspectives. In other words, the measures were
not neatly tied back to the theoretical perspective they purported to represent.

The approximation of another critical test unfortunately suffers from the
same kinds of ambiguities. It began with the commonplace observation that in
modern industrial societies, increasing amounts of education are required as
prerequisites for employment. The investigator then asked why this happens
and posed two alternative explanations. The first, a technical-functional
theory, was tied to Davis and Moore. Increased skill is required, according to
this view, because of increasingly complex technology. The alternative, a
conflict theory, views the increasing prerequisites as the arbitrary impositions
of organized groups who are trying to dominate the job field.[31]

These two perspectives were then rather loosely assessed in light of
survey data that had been previously gathered for a variety of purposes. For
example, an examination of historical changes in the skill required to perform
a number of jobs shows only very small increments. Educational levels associated with these same jobs appear to have increased in excess of the
increase in actual skill required. This suggests to the investigator (i.e., Collins) that the historical trend is better explained by the conflict than the
functional perspective. Actually, findings of this type do cast doubts on the
functional view, though they do not provide clear support for the conflict

---

[30]See Huaco, "The Functionalist Theory of Stratification."
[31]Randall Collins, "Functional and Conflict Theories of Educational Stratification,"
American Sociological Review, 36 (1971).

view. In other words, while questioning the applicability of the functional explanation, they do not really demonstrate what other perspective might be more applicable. Similarly, the investigator questions whether formal education does in fact provide job skills, as implied by the functional theory. Apparently not, is the tentative answer, as indicated by other survey data, which indicate that level of education and productivity are not positively related. Most learning of skills appears to occur on the job.

Thus, the study concludes, the demands of occupational positions are not fixed by functional requirements, but appear subject to negotiation or bargaining among organized groups. There is, as previously noted, little direct evidence for this conclusion; however, the questions that are posed and the data that are assessed do suggest some real limitations to the functional perspective.

Conclusion

In the preceding section of this chapter we have examined a large and diverse assortment of studies that have attempted to assess aspects of the functional theory of stratification. It has not been a complete review, but I believe it to be representative of all such studies that have been published. Therefore, based upon the evidence, the time has come to attempt to offer some conclusions, though the methodological problems inherent in the studies will necessitate that all conclusions be regarded as extremely tentative.

A central assertion of Davis and Moore (and to a lesser extent, Parsons and Warner) is that social positions differ in their intrinsic importance to society, and these differences produce differences in reward. In the two studies, which were based upon people's perceptions, the results indicated that functional importance was associated with reward, but only to a small degree; much less than the talent required, and much less than the theory implies. In the studies where functional importance to society was inferred without resorting to perceptual measures, the results were substantially more congruent with the theory. The latter studies, however, did not systematically deal with talent required, the other half of the Davis-Moore equation. Therefore, two critical issues become evident: the validity of perceptual measures, and the relationship between functional importance and talent required. Let us re-examine each in turn.

The validity of perceptual measures rises or falls on the key assumption that: "highly important positions are recognized as such"; or at least, on the assumption that perceptions provide an accurate "first approximation."[32] I am dubious of either assumption, even though they have been widely main-

[32]Lopreato and Lewis, "An Analysis of Variables," p. 302.

tained, and even though they are congruent with Parsons's earlier theoretical formulation. My negative stance rests on the belief that prestige is such a salient feature of occupations that respondents find it impossible to differentiate between it and other attributes of occupations, such as functional importance. This conclusion is very similar to that reached by Lopreato and Lewis in their study of perceptions. Further, both historical and cross-national studies have had little difficulty obtaining consistent results when focusing upon ratings, which also suggests that prestige may be *the* perceived attribute of occupations. Therefore, I am inclined to believe that perceptual measures of functional importance ought to be disregarded. This also appears to be Huaco's conclusion when he asks, in reference to Stinchcombe, what people's perceptions "have to do with differential functional importance?"[33]

The non-perceptual studies do support the contention that rewards are tied to functional importance, even though they do not demonstrate causality, and I am inclined to weigh this evidence more heavily. However, these studies have not shown that functional importance operates independently of the talent required for the position. The Davis and Moore formulation, it will be recalled, is ambiguous on this point. The only direct evidence is presented in the perceptual studies, and such studies are of no help because even if we were to accept their validity, they offer opposite conclusions. Lopreato and Lewis show that much of the apparent relationship between functional importance and reward is actually mediated by the relationship between functional importance and skill required. Grandjean and Brown report, however, that importance and education required are independent of each other in producing rewards. It must also be pointed out, though, that to make the studies comparable in this regard I am equating education and skill, and this may be unwarranted.

It seems clear that in the more than thirty years since its presentation, the functional theory is yet to produce anything resembling definite evidence in support of it. The most encouraging sign, from an empirical standpoint, is that in recent years there have been vastly increasing numbers of attempts. Whether or not any theory stimulated such efforts is one important criterion by which theories are scientifically evaluated, and the functional theory shows promise of earning passing grades in this regard. However, it also seems clear, at least to me, that the end result of all the studies will reinforce the need to develop other theories of stratification along with the functional view. The latter is not sufficiently sensitive to the strangulation of talent issues and the role of power in setting job requirements. These are, of course, familiar problems that we have previously encountered in prior chapters, which were devoted to an examination of more general and global functional theories.

[33] Huaco, "The Functionalist Theory of Stratification," p. 234.

5

deviance

The functional approach to the study of deviance lacks the specificity and concreteness associated with the functional theory of stratification. Both the concepts and the propositions relating to deviance are more abstract. Therefore, it will not be possible in this chapter to review equally specific tests of specific hypotheses.

At a theoretical level, the functional approaches to deviance and stratification possess similarly high degrees of convergence, but this convergence has occurred in different ways. During the 1940s, consistent views of stratification and its functions were formulated by a number of writers. The similarities among contemporary functional theories of deviance are more the result of their historical continuity with the earlier and more general statements of the functionalist perspective. Durkheim in particular has very strongly influenced the viewpoints of modern functionalists and he has correspondingly been the central figure in more recent polemics.

Durkheim's contributions include both a theoretical view of deviance as normal and functional, and a methodological model that utilizes official statistics in deductive tests of the theory. The theory and method are separable and will be analyzed separately in this chapter, though they have historically become intimately interwoven and together they provide the distinctive hallmarks of the functional approach to deviance.

Because of their enormous influence, Durkheim's writings provide the skeleton around which this chapter is organized. We will begin with an overview of his position, and then focus upon its major theoretical and methodological aspects. A variety of recent studies will be reviewed, primarily to illustrate the kinds of insights that have been stimulated by Durkheim's contributions. In reviewing the criticisms of the perspective, and the important disagreements among functionalists, Durkheim will continue to be viewed as the central figure.

Durkheim's View

Durkheim's discussion of deviance poses as a central question whether or not criminal behavior should be considered "pathological."[1] His model of society in this case was explicitly physiological and the question was whether crime in society could be seen as analogous to disease in a biological organism. At first glance he believed many people were inclined to regard it as pathological in this sense; but, Durkheim contended, any kind of behavior is normal rather than pathological if: 1) it is usually associated with a given type of society, and 2) its rate of occurence is within certain limits.[2]

The second point above emphasizes a statistical view of normality. Crime, or deviance more generally, is abnormal only if its rate of occurence deviates from some normal rate, one which is a statistical average for all societies of a given type. This introduces Durkheim's first point, namely, that the forms which deviance takes are intimately related to the social organizations of the societies in which they occur. Thus, behavior can be considered abnormal or pathological if it deviates statistically or if it deviates in form.

Both of Durkheim's points can be illustrated in his classification and analysis of suicide. One type, called "altruistic suicide," involves an overcommitment to the collective life. As an illustration, consider an adolescent male in a primitive society who is thought by others to be cowardly. This evaluation might lead the youth to believe that he was a public disgrace and an embarassment to his tribe, and he might respond by deliberately taking his own life. This case provides a good example of an altruistic suicide, and

---

[1]Crime is not typically equated with deviance. The former is conventionally treated as a specific and limited case of the latter, a type of deviance that is covered by criminal statutes. However, Durkheim's discussion of crime was sufficiently encompassing for it to be generalized to deviance.

[2]Emile Durkheim, *Rules of Sociological Method* (New York: Free Press, 1950).

Durkheim argued that suicides of this type were likely to occur, at a specific rate, in societies where individuality is strongly subordinated to collective life. Associated with this de-emphasis upon individuality are societies in which the division of labor is minimally differentiated.[3]

A certain rate of altruistic suicide was expected to occur as part of the normal course of events in such homogeneous societies. Altruistic suicides would be considered pathological only if their rate of occurrence deviated from the rates typically observed in societies of the same type. However, Durkheim also conceptualized other types of suicide, and their existence to virtually any degree in such primitive societies could be considered pathological. Anomic suicide, for example, occurs in response to rapid social changes when the norms that restrain individual aspirations lose their regulatory force. This results in frustrated ambitions, discontent and despair for individuals, and a high rate of anomic suicide in the society. While there is a normal rate of anomic suicide in modern (i.e., highly differentiated) societies, their occurrence in primitive (i.e., non-differentiated) societies would be considered pathological.[4]

In addition to emphasizing rates and forms, Durkheim also stressed that crime, or deviance, had positive consequences (i.e., was functional) for society. This conviction stemmed from his view, discussed in Chapter 2, that all social facts tend to contribute to the "harmony of society." And criminal or deviant behaviors, if they occurred within certain limits, were regarded as normal social facts. "In the first place crime is normal because a society exempt from it is utterly impossible."[5] In various writings, Durkheim attributed a number of specifically positive consequences to deviance. One was a clarification of social norms. The existence of crime indicates that there is a degree of flexibility among the collective sentiments within a society. Crime can lead to a crystallization of these sentiments and may even help to determine the direction in which public morality will change. Deviance can also make people more aware of the values they share, thereby contributing to social solidarity. Therefore, Durkheim concludes, crime "must no longer be conceived as an evil that cannot be too much suppressed."[6]

It must be recalled, however, that Durkheim was analyzing normality from the standpoint of the society. Hence, what applies to crime does not necessarily apply to the criminal. The person who commits deviant acts may or may not be normal, psychologically. Furthermore, there is no reason to assume that social utility constitutes an important motive for the individual who commits the deviant act. As a general procedural matter, Durkheim

---

[3]Emile Durkheim, *Suicide* (New York: Free Press, 1951).

[4]The third type, "egoistic suicide," is the opposite of altruistic. It occurs when individuality is emphasized to the point of producing egoism.

[5]Durkheim, *Rules of Sociological Method* (New York: Free Press, 1950), p. 70.

[6]*Ibid.*, p. 72.

rejected efforts to infer motives. He claimed that no sociological *uniformities* could be built from them when studying suicide because the motives offered were likely to be as varied as the actual number of suicides. Thus, motives were not worth bothering with, in a pragmatic sense. Even more fundamentally, though, motives do not constitute a social fact and all social facts (such as suicide rates) must be explained by other social facts.

Many of Durkheim's views on social pathology, anomie, and crime were presented in his study of suicide. This study was also methodologically very innovative, particularly for turn of the century sociology. To test his theories, Durkheim collated extensive comparative and historical data. This included suicide rates starting from 1841 and continuing through to 1872 in France, Denmark, and a number of other European countries. He then compared suicide rates to overall mortality rates and examined variations in suicide rates within countries according to religious affiliations. In this ambitious study he provided a model of how ensuing students of deviance might test theories utilizing "real world" data.

In terms of subsequent influence upon sociology, Durkheim's major impact upon the study of deviance can be summarized as follows:

1) A view of deviance (within limits) as normal and functional for a society.
2) An emphasis upon the inter-connections between social organization and forms of deviant expression.
3) The extensive utilization of "official" (i.e., government and census-type) data to test theories of deviant behavior.

In the following pages we will examine each of the above contributions, in turn. Recent studies will be reviewed as illustrations of the theoretical insights that Durkheim's writings helped to stimulate. In relation to each contribution we will examine the disagreements among functionalists and the criticisms of others that have been directed at the viewpoint.

The Function of Deviance

One of the most intriguing efforts to show the positive consequences of deviance is provided by Erikson's award-winning study of the sixteenth and seventeenth century Puritans of New England.[7] In order to study the role of deviance in Puritan communities, he examined official statistics, such as court records; personal documents, such as diaries; and a variety of other sources.

Judged by contemporary standards, the Puritans placed tremendous stress upon conventional morals. They created communities in which the laws

[7]Kai T. Erikson, *Wayward Puritans* (New York: Wiley, 1966).

were heavily infused with fundamentalist religious values. Their persecution of Quakers, their witchcraft hysteria, and other collective movements were viewed by Erikson as continuing attempts to define and re-define the community's social "boundaries"; that is, to set the limits upon compliance and deviance, and to maintain the distinctiveness of the community. Movement of these boundaries, in effect, created "crime waves," as persons whose behavior was formerly in the range of the acceptable were suddenly deviant. The branding of heretics and witches that followed sustained community norms and solidarity. Thus, movements of the boundaries can be viewed as highly functional to the maintenance of a distinctive community. An interesting example is provided by the trials of Anne Hutchinson, who was ultimately excommunicated from both the church and the community (though the distinction between the two was blurred).

In the 1630s, the Hutchinson's Boston home was a center of lively theological discussion. Community interest in Mrs. Hutchinson's biblical scholarship and viewpoints came to exceed interest in the minister's "official" sermons. And because she was critical of the political-religious ruling body, the battle lines were soon drawn. The fact that the dissent was spurred by a woman—whose rightful place was believed to involve housework rather than scholoarship—made her position still more tenuous. Her position was threatening to the establishment, Erikson notes, because it denied the ministers' ability to utilize the covenant of grace as a political instrument. Thus, while Ann Hutchinson and her followers appeared to be dissenting on theological grounds, she was initially charged with sedition rather than heresy.

The civil trial was a sham, even by seventeenth century standards. Her claim that her position was based on the revealed word of God was viewed by the court as a "devilish delusion." After polling the court, Governor Winthrop declared that, "the danger of her course among us . . . is not to be suffered . . . Mrs. Hutchinson is unfit for our society."[8] After a four month imprisonment she was then tried by the Church, and "delivered over to Satan." It was another ritual trial whose outcome was never in doubt.

Throughout the civil trial, Mrs. Hutchinson repeatedly requested to know the charges that were placed against her. While all members of the court acted as though the answer was self-evident, no specific charges were ever given. She could only be told that her conduct could "not be suffered" because there existed in the colonies no names for her crimes. However, she had to be found guilty so that she could be banished, Erikson concludes, because it was the only way to protect the community's boundaries.

In this study of Puritan life, Erikson explicitly began with Durkheim's assumption that, "deviant forms of behavior are a natural and even beneficial part of social life." What will be regarded as deviant, in any community, depends upon what is considered "dangerous" or "embarrassing" or "irritat-

8*Ibid.*, p. 99.

ing" to the "people of a group"; and this will change with the "shifting mood of the community." Communities maintain boundaries in order to preserve their "cultural integrity" or distinctiveness. From this perspective religious or civil trials, courts martial, and even psychiatric confinements can be viewed as ways of drawing boundary lines. Thus, deviant behavior is socially generated and, within limits, provides an important, albeit generally unrecognized, means for preserving the special character and the stability of social life.[9]

A similar perspective is provided in Merton's analysis of big-city political machines. In a number of ways, they appear to involve practices that directly contradict public morality, such as patronage and bribery. Political machines, then, might be considered bad or undesirable. This evaluation, however, looks only at the manifest functions of machines. It disregards the latent—generally unrecognized and unintended—consequences. To apply a functionalist perspective, Merton continues, the sheer persistence of any social pattern or structure must be taken as an indication that it is probably "satisfying basic latent functions."[10]

He then proceeds to analyze the larger social context in which the machines operate. American political power is fragmented and dispersed, he notes, not only at the federal level, but at the local level as well. This creates cracks or voids in the system; situations in which individuals are confronted by a bewildering array of official channels. In this context the machine provides an effective, all-purpose source of assistance. For large businesses this entails needs for licenses, city contracts, and the like. For minority groups it entails information about and access to hospitals, legal aid, immigration authorities, etc. And for certain subordinated groups, whose access to conventional means of success is blocked, the machine itself can provide an alternative route to social mobility.

Merton, it should be explicitly noted, is meticulous about specifying who benefits from the various structures that make latent contributions. This attempt to be concrete follows from his criticism of Durkheim's amorphism in viewing society as the beneficiary. What is functional for some segments of society, he argues, may be dysfunctional for others.

Despite this stated specificity, Merton's case studies—political machines, for example—tend to suggest that latent functions serve highly diverse needs in highly diverse groups, from segregated minorities to legitimate businesses. Even though such segments of the society appear to be poles apart, the fact that they are simultaneously served by the same ongoing practices implies that it may be possible to generalize about functions for society. Thus, with respect to political machines as crack-fillers, Merton concludes,

[9]Among later functionalists, the view of deviance as functional is most conspicuously de-emphasized by Parsons. He stresses a view of deviance as disruptive and regards it as "immoral," if not pathological. See the discussion of Parsons in Chapters 2 and 3.

[10]Robert K. Merton, *Social Theory and Social Structure* (New York: Free Press, 1949), p. 71.

"the several subgroups in the large society are 'integrated' . . . by the centralizing structure."[11]

Merton's stated concern with specifying who benefits had been anticipatory of what was to become a major criticism of Durkheim's view, that deviance was functional "for the society." The persons labelled as criminals and punished for their acts are, after all, parts of society too; and, the critics contend, it is difficult to see how they benefit. The general issue involved here is one of of sensitivity to questions of power. By emphasizing society as the beneficiary, the critics argue, Durkheimian functionalism is led to disregard differences in the social power of various segments of the society. Neither Erikson, Merton, nor others following in this theoretical tradition are sensitive to ways in which the power of certain groups can determine whether aspects of their life style will be considered deviant. One source of this insensitivity, according to Gouldner, was the preoccupation of Durkheim and the anthropological functionalists with primitive societies that were without complex political systems. This focus made it easy for functionalism to disregard "political relevance."[12]

In order to grasp this theoretical power void, let us pause for a moment, recognizing that all the functionalists tend to agree that the labeling of certain acts as deviant has an arbitrary element involved in it. It is capricious in the sense that the behavior that is stigmatized is not *inherently* deviant. At other times or places the very same behavior is lauded. Ann Hutchinson's defiance of the establishment and her skill in debating, for example, would have earned her great esteem on American campuses in the late 1960s. In other segments of the same society, however, this defiance—especially by a female—would have been considered a violation of normative standards; but transgressing the boundaries in this manner would not have likely led to her excommunication in the literal sense.

Becker makes this general point when he argues that factions within a society are likely to hold different definitions of what is acceptable behavior. These differences ultimately become subject to political conflicts in which it is decided whose "rules are to be enforced" and whose behavior is to be "regarded as deviant."[13] The outcome of this political process involves definite winners and losers. Whosever definition of acceptable behavior is rejected will have aspects of his or her life style prosecuted as a result. An interesting example of what "winners" gain is provided by Becker's analysis of the Federal Bureau of Narcotic's role in urging passage of the Marijuana Act in 1937. He views the Bureau as having been a "moral entrepeneur" in advocating passage of the Act, and as having its own stature and resources increased as a result of its enactment.

[11]*Ibid.*, p. 81.

[12]Alvin W. Gouldner, *The Coming Crises of Western Sociology* (New York: Basic Books, 1970), p. 130.

[13]Howard Becker, *Outsiders* (New York: Free Press, 1963), p. 7.

In sum, the Durkheimian view that deviance is functional has prompted a number of conceptual breakthroughs in the sociological study of deviance. Most importantly perhaps, it provided an outlook with which sociological views of deviance could be wrenched apart from "public morality." This break was extremely useful because it permitted sociological theory to develop beyond the vagueness and inconsistency of common sense viewpoints.[14] In addition, Durkheim's functional perspective also led to the formulation of useful "sensitizing concepts," such as Erikson's notion of boundary maintenance. Such sensitizing concepts are difficult to test empirically, and their greatest utility is probably with an ad hoc, historical analysis. However, their sensitizing qualities make them an important contribution to the conceptual "tool kit" of the sociologist.[15]

On the other hand, the view of society as the beneficiary is ensnarled in all of the conceptual and methodological difficulties that have previously been seen to be associated with the conception of society as "a thing apart." In particular, this functional view tends to underplay the diversity of life styles in a society. It also de-emphasizes the degree to which power is the mechanism that allows different segments of a society to impose their definitions of appropriateness in the labeling of deviance.

Where one stands on these theoretical issues has profound consequences for how concrete events will be interpreted. As an illustration, let us reconsider the trials of Ann Hutchinson. Her trials can be viewed as a community's attempt to establish normative boundaries. The magistrates and ministers are correspondingly seen to be acting as agents of the community. Or, the same trials can be viewed as an exercise in power. Those segments of society whose ox was being gored, flexed their muscles—i.e., utilized their control over conventional structures (e.g., courts)—in order to maintain their established advantages and discourage future attempts to erode their power and privileges.

Congruence of Forms

A second major emphasis that is derived from Durkheim's view lies in the expectation that there will be a strong relationship between the form or rate with which deviance occurs and the nature of social organization. Congruency was a general proposition of Durkheim's, but one which he most extensively applied in an evolutionary perspective. Specifically, undifferentiated societies, characterized by homogeneity and mechanical solidarity, would have distinctive types of both deviance and law. The differences Durkheim postulated between these types of societies have been subjected to considera-

---

[14]See Merton, *Social Theory and Social Structure*.

[15]An overview of functionalism's contributions, written from a critical viewpoint, is provided by, David Matza, *Becoming Deviant* (Englewood Cliffs, N.J.: Prentice-Hall, Inc. 1969).

ble criticism, especially from other functionalists, and most notably by Malinowski. The general assumption of congruency has become entangled with a different set of problems, although both sets of criticisms lead in similar directions. For our purposes here, however, we will first consider the general assumption by itself.

A good example of the type of inquiry Durkheim's view helped to foster is provided by Kingsley Davis's analysis of prostitution. (This is the same Davis whose theory of stratification was examined in the preceding chapter.) In his view, the prevalence of prostitution is a function of the degree of sex-role specificity in the larger society. According to this theory, when women's roles are defined in a highly traditional manner, men will seek sexual gratification outside the home. Here, a specialized group of women will emerge for this purpose; for example, prostitutes and geishas. By contrast, when women's roles are defined in less traditional ways, their outside-the-home experiences make them more "interesting," and conventional norms that limit appropriate sexual activities are apt to be less rigid. As a result, prostitution will flourish to a substantially lesser degree.[16]

Davis explicitly interprets this relationship as due to the relatively constant needs of males for sexual gratifications. This explanation is highly unsatisfactory because it minimizes the role of the female's sexual needs and the role of culture in shaping drives. Besides, more implicitly, Davis offers a better interpretation when he views the sex-role and prostitution association in relation to boundary maintenance. Thus, when feminine roles are traditionally defined, the conventional sexual boundaries of women are narrowly circumscribed. Transgressions are unambiguous and the normative structure itself leads to the creation of special categories of "immoral" women. When sex roles are defined less traditionally, by contrast, the same types of behavior on the part of the female are evaluated in a different social context and do not lead to labeling women as prostitutes.

Support for Davis's hypothesis can be gleaned from a number of sources. The results of self-report surveys from Kinsey (1948) through Hunt (1973), for example, indicate that male-reported incidence of contact with prostitutes has declined markedly. Over this same twenty-five year period, women's roles in the United States appear to have become much less traditional. The same expected relationship between sex-role "modernization" and the prevalence of geishas appears to have occurred in Japan also.[17]

[16]Kingsley Davis, "The Sociology of Prostitution," *American Sociological Review,* 2 (1937). A revised presentation is in Chapter 7 of Robert K. Merton and Robert A. Nisbet, eds., *Contemporary Social Problems* (New York: Harcourt Brace, 1966).

[17]These and other supportive findings are presented in Carol Leonard, "Prostitution and Society" (unpublished Ph.D. dissertation, Department of Sociology, Syracuse University, 1977). A similar analysis of male homosexuality in Greece is provided by Gouldner who relates it to female sex roles and to its congruence with the competitive pressures of the Greek contest system. See Alvin W. Gouldner, *Enter Plato* (New York: Basic Books, 1965).

Evidence of this type does not constitute proof in a causal sense because of methodological problems associated with these analyses. They are specific to each study and need not concern us here. Of more pressing concern is the question of whether functional analyses, such as Davis's, actually explain the phenomenon in question. Has prostitution, for example, been explained when it is related to sex roles? From the critics' standpoint, Davis may have said something about sex roles, but the presumed object of inquiry, prostitution, seems to remain unanalyzed.[18] We will return to the issue of whether indicating congruence constitutes an explanation. First, however, let us examine Durkheim's evolutionary analysis of deviance and social organization.

Durkheim's evolutionary view focused upon the division of labor in two polar types of societies, simple (i.e., undifferentiated) and complex.[19] Corresponding with the simple form: deviations trigger an immediate response from the "general population"; no specialized agents, such as police or courts, are present to respond; and the overlap between the collective conscience and law is sufficiently complete that almost all crimes can be viewed as moral violations. The culprit is punished as a signal to others, to prevent future occurences, and also to drain off the anger that the violation aroused.

With increased differentiation, the collective conscience shrinks and comes to overlap less with law. (Does anyone feel moral indignation when somebody cheats on their income tax?) Law correspondingly functions to make amends; to re-integrate the differentiated parts. (Thus, income tax cheaters are liable to fines rather than public hanging.)

Associated with these differences between societies, Durkheim distinguished between two types of law, calling them "repressive" in simple societies and "restitutive" in societies that are complex. Repressive law was considered as the only type present in simple societies. Complex societies possessed both types of law, restitutive being added "on top of" repressive, and tending to be the predominant type. Radcliffe-Brown emphasized a similar distinction, but modified the terminology. He differentiated between "custom" (simple) and "law" (complex), the latter involving specific sanctions.

The distinction between types of law, or between custom and law, has been severely criticized by Malinowski for exaggerating differences between societies. In any society, he argues, the "binding forces" are longstanding mutual obligations. If an individual—whether Melanesian native or civilized businessman—fails to meet his obligations (i.e., deviates) the "weapon for coercion" is clear: reciprocity. This can involve re-payment in kind, or the withholding of expected services. Everyone consciously realizes this, Malinowski continues, and their fear or vanity keep them in line most of the

---

[18]See, for example, Joan Smith, "The Failure of Functionalism," *Philosophy of Social Science,* 5 (1975).

[19]The inadequacies of this dichotomy are discussed in Theodore D. Kemper, "The Division of Labor," *American Sociological Review,* 37 (1972).

time. Of course, their laziness or ambitions will sometimes lead them to try to escape from obligations. Thus, Malinowski argues that processes of compliance and deviation are very similar in all societies. They are, first of all, more dependent on psychological dispositions than Durkheim realized, and secondly, people everywhere are kept in line by the same pressures of reciprocal obligations. This he termed, "the functional theory of effective custom."[20]

Another related and major disagreement between Durkheim and Malinowski involves what the latter terms a view of "automatic submission." Durkheim's underlying assumption, according to Malinowski, is that in simple societies the individual is totally dominated by the group; that is, he acquiesces like a passive robot. Malinowski seriously questions this. Does anyone, he asks, either civilized or savage, carry out obligations in this automatic manner? No, is his emphatic answer. There are always complex inducements that are both psychological and social in nature. In addition, systems of law and custom contain contradictory directives that conflict with each other, forcing individuals to make choices and thereby guaranteeing that, from one jurisdictional view or another, some behaviors are certain to be deviant. What will then happen, he insists, depends upon concrete circumstances. Durkheim's emphasis upon automatic submission, Malinowski concludes, misses all of these nuances.[21] For example, Durkheim viewed altruistic suicide as occurring in an automatic, non-problematic sequence, in simple societies.

In order to illustrate his argument, Malinowski describes the events that led to the suicide of a young Trobriand male. He had violated the community's incest taboo by having sexual relations with his maternal first cousin. The natives express horror at such acts, in the abstract. According to the ideals of native law, such acts cause the world to reverberate and invite disease and death to plague the community. In reality, however, even the people who knew what had transpired were not very outraged or afraid. The Trobrianders had a well-established "system of evasion," consisting of various magical practices that could have been utilized to effectively pacify the outraged gods. Thus, there need not have been any robot-like adherence to tradition resulting in compulsory suicide to appease a deity.

The situation changed dramatically, however, when a rival for the youth's cousin publicly insulted the youth with the charge of incest. Then ideal sentiments became prominent. Feeling intolerable shame the youth climbed to the top of a sixty-foot palm tree and leaped to an instant death.

Malinowski emphasizes the discrepency between actual sequences of

[20]See the introduction by Bronislaw Malinowski in H. Ian Hogbin, *Law and Order in Polynesia* (Hamden, Conn.: Shoe String Press, 1961). (Originally published in 1934.)

[21]Bronislaw Malinowski, *Crime and Custom in Savage Society* (London: Routledge and Kegan Paul, 1926).

events, in which resultant outcomes are variable and problematic, and Durkheim's view of rigid and automatic compliance. Given this discrepency, he questions why anyone believes the "fictitious," overly simplified ideal. He proceeds to reply to his own question with an epistemologically provocative answer.

> "When the native is asked what he would do in such and such a case, he answers what he *should* do . . . it costs him nothing to retail the Ideal. . . . The other side, the natural, impulsive code of conduct, the evasions, the compromises . . . are revealed only to the field-worker, who observes native life directly . . . truth is a combination of both versions . . . [which are] . . . futile simplifications of a very complicated state of things.[22]

Thus, Malinowski's main criticism of the congruence of forms thesis is its naivety; a naivety that results from applying an abstract perspective in a guileless manner. This in turn results in explanations that do not fit the concrete phenomenon. It is essentially the same argument we encountered previously in regard to Davis's theory of prostitution; namely, that the functionalists offer "non-explanations." At the heart of this disagreement are different conceptions of what constitutes an explanation, which, in turn, are associated with different theoretical strategies.

For a generalizing science, such as sociology, the mandate is clear: to develop generalizable theories. This requires abstraction, or the conceptualizing of phenomena apart from their concrete representations. We all engage in a similar process when we conceive of furniture rather than tables, chairs, etc. The functionalists have approached this task by attempting deductively to apply general theories of deviance to specific forms (i.e., crime, suicide, etc.). By analogy, one could begin with an abstract conception of furniture, its characteristics, its utilities, and so on. In applying this abstract conception, one's description of chairs, tables, or sofas would be lacking in detail. And a major criticism of the functional theories is precisely that they lack specificity, richness, and attention to detail.[23]

One of the most influential discussions of this issue was presented by the German sociologist, Max Weber, who noted that while sociological analysis abstracts from reality, it also helps us to understand reality. This is true, he continued, even though the correspondence between the abstract and its concrete representations must always be less than perfect.[24] However, while Weber's view of sociology as an abstract and general science is compatible with some features of functional theories, his emphasis upon subjective

---

[22]*Ibid.*, p. 120-21.

[23]See, for example, Matza, *Becoming Deviant;* and Albert Szymanski, "Malinowski, Marx and Functionalism," *The Insurgent Sociologist,* 2 (1972).

[24]Max Weber, *The Theory of Social and Economic Organization* (New York: Oxford, 1947), Part I.

meanings is not consistent with other features. Any explanation, Weber argued, must adequately account for the subjective interpretations of participants. Statistical rates, therefore, belong in sociological generalizations only when they can be regarded as, "manifestations of the understandable subjective meaning of a course of social action."[25]

This emphasis upon subjective interpretations places limits on generalizations. Thus, to pursue our earlier analogy, tables, chairs, and sofas may be viewed abstractly as furniture because of the shared social meanings regarding each of these concrete things. (Unlike people, though, objects of furniture have no shared meanings among themselves, which makes the example much simpler than the typical sociological problem.) The critics of functionalism emphasize adequacy at the level of meaning when they decry a lack of detail, richness, and specificity. An inductive approach is called for, they contend, in which theory is built from microscopic examination of concrete actions. There is, then, at least some agreement on what constitutes a desirable goal: general theory. The issue is how to get there.[26]

Official Statistics

One direction pursued by the debate about the adequacy of meanings involves the functionalist tendency to utilize official statistics in deductively testing theories. Durkheim's study of suicide, which methodologically provided a prototype for ensuing functionalists, utilized a variety of government statistics, many of which had been used by others in prior studies of suicide. By the time Durkheim used them there were already questions regarding the validity of these statistics and Durkheim was aware of them. In addition, Durkheim's own definition of suicide was not fully developed and there is the question of whether his definition corresponded with those of designated functionaries whose decisions determined which deaths were to be counted as official suicides.[27]

Despite his disdain for individual motives, Durkheim was forced to confront motivational and cognitive elements in order to define suicide. Specifically, he defined suicide as death that results from the deliberate and knowledgeable acts of individuals; that is, the individual acted (e.g., leaped) or failed to act (e.g., avoided a train) deliberately, and the individual recognized a reasonable likelihood of death resulting from the act or its omission. Thus, Durkheim offered a general, abstract meaning—for both the contem-

[25]*Ibid.*, p. 100.

[26]See Paul G. Schervish, "The Labeling Perspective," *The American Sociologist,* 8 (1973).

[27]For an exhaustive and insightful analysis of suicide rates and social meanings, see Jack D. Douglas, *The Social Meanings of Suicide* (Princeton, N.J.: Princeton University Press, 1967).

plator of the act and for others—and he did not regard this social meaning of suicide as problematic. However, he did not indicate how such meanings could be deduced from concrete and observable phenomena. He simply assumed, in an ad hoc way, that the observable phenomena were indicative of the theoretically developed categories.

Beginning with the observed fact that a death has occurred, how is it possible, following Durkheim, to determine intent?[28] How is it also possible to know what inferred meanings will be utilized by officials to decide whether it is to be classified as a suicide?[29] The only way, Douglas states, is by, "studying the specific meanings . . . as the individuals involved construct them" and then work "upward to abstractions" about more general meanings.[30] However, when the official rates are uncritically taken as givens, these problems of meaning are swept under the rug.

To this point, we have discussed official records with respect to suicide, but a moment's reflection makes it apparent that very similar issues are involved in the study of prostitution, crime or virtually any other type of deviant behavior. If it could safely be assumed that the *true* distribution of any form of deviant behavior was representatively reflected in the official rates, then they could be used as unbiased indicators. However, there is no true rate that exists apart from social meanings. Further, surveys in which people provide self-reports on a variety of criminal acts often show marked differences between "hidden" and "official" offenders; differences in prevalence, frequency, and visibility.[31]

It should also be realized that social scientists inadvertantly contribute to the generation of official statistics that may support their theories for the wrong reasons. For example, if a lower class milieu is reported to produce excessively high rates of male juvenile delinquency, the findings of researchers may fortify the already existing class biases of police and juvenile judges. Their biases may lead them to view petty acts (such as vandalism) as indicating serious delinquency if committed by a lower class youth, but dismiss it as "boys being boys" when middle class youths are the offenders. This is the stuff out of which official rates are created. (Note that the social meanings remain *implicit*.) Thus, the biases of officials may produce a distorted official record, which sociologists take as data, reporting findings that reinforce the original biases. This leads to the perpetuation of biases, and future studies have little difficulty replicating earlier ones.

[28]On this point, see Jerry Jacobs, *Adolescent Suicide* (New York: Wiley Interscience, 1971).

[29]On this point, see George Simpson, "Methodological Problems in Determining the Aetiology of Suicide," *American Sociological Review,* 15 (1950).

[30]Douglas, *The Social Meanings of Suicide,* p. 253.

[31]For a review of these studies, see Albert K. Cohen, *Deviance and Control* (Englewood Cliffs, N. J.: Prentice-Hall, 1966), pp. 24-29.

Even more serious, perhaps, are the theoretical influences operating upon social scientists, which lead to unconscious distortions of the data that they themselves generate. Consider as illustrative the role of the ethnographer in studying a primitive society. On the one hand, the researcher is presumably a neutral observer who records and analyzes events. On the other hand, the researcher must utilize either private meanings, those of natives, or, more likely, some combination of both in order to classify behaviors as constituting events. Should a death be considered a suicide? Has a crime been committed? Thus, the researcher is not only an observer, but also plays what is in effect the role of official. The theoretical orientation with which the ethnographer enters the field may influence resultant perceptions and produce distorted records in exactly the same manner as with presumably less sophisticated officials, and with the same result: an official record is constructed that contains biases inherently congenial to a theoretical perspective. (Ethnographic descriptions of primitive societies, it will be recalled, have provided an important source of data for the testing of functionalist hypotheses.)

An interesting illustration of this process is provided by Carroll's analysis of differences in ethnographic reports that were published in what he terms the functionalist period and the pre-functionalist period.[32] While recognizing that there is no exact date that separates the two periods, for purposes of the inquiry he arbitrarily selected 1930. Even though the writings of Durkheim, Malinowski, and Radcliffe-Brown began substantially before 1930, Carroll states that around that time their combined influences created a functionalist era that has continued to the present time. Despite their own differences, he argues, they shared a common emphasis upon functional unity: a belief in the integration of social systems. As a consequence of this perspective, he concludes, the ethnographers working in the functionalist era may have been led to underestimate the incidence of deviant behavior. Certainly the post-1930 influence of Parsons's functional synthesis has been criticized for its emphasis upon stability and order.

Parsons generally regarded deviance as very sporadic in occurence. He allowed for some rate of occurence, even in a highly integrated social system, but its magnitude, though unspecified, was apparently very low. Two of the major sources of deviance were identified by Parsons as involving conflicting values and socialization failures. With respect to the former, contradictory expectations were viewed as simultaneously coexisting in any society. They do not typically cause disturbances in the social system because the conditions under which they apply are kept separate; but some individuals or groups can still be expected to trip over the inherent contradictions and behave inappropriately. Failures of the socialization process involved, for example, children

[32]Michael P. Carroll, "The Effects of the Functionalist Paradigm Upon the Perception of Ethnographic Data," *Philosophy of Social Science,* 4 (1974).

not being taught to anticipate future role expectations or being raised in a manner that created needs whose expression ran counter to prevailing standards.[33]

Thus, deviance entered the picture at critical points of connection between the social system and the cultural system (i.e., contradictory values) or between it and the personality system (i.e., socialization failures). Correspondingly, Parsons emphasized the dysfunctions of deviance. Even though he recognized that deviance could be the source of important social changes (e.g., the clarification of conflicting values), he stressed the way in which it disrupted the integration of the social system. Thus, the post-1930 functionalist period, as described by Carroll, is predominantly a Parsonian-influenced era.

In order to document the possible existence of this functional bias, Carroll classified several hundred ethnographic reports according to their time of publication. The functionalist conception of system integration should have led, he hypothesizes, to lower perceived rates of murder, suicide, crime, and other deviant acts after 1930. His findings tend to support the hypothesis, showing lower post-1930 incidences of reported practices that could be regarded as indices of social instability. For example, estimated suicide rates were offered for thirty-five societies, fifteen before 1930 and twenty after. In the pre-functionalist period, 40 percent of the studies reported low rates of suicide compared to 60 percent in the functionalist period. Differences of similar magnitude are reported for all seven indices of instability reported by Carroll.

It is possible, of course, to dismiss these findings. The differences are not very large, typically, and they might be due to chance. Utilizing publication dates may have led to mis-classifications, for example, based upon varying time lags between time of research and time of publication. It is also possible that ethnographic studies became methodologically more sophisticated and more accurate in estimating rates, or that the societies studied actually underwent changes in the reported directions. Thus, the differences may be spurious; but even if they are not, they may be attributable to a variety of factors other than a functionalist influence upon ethnographers. Nevertheless, Carroll's study clearly illustrates the kinds of subtle forces that may be operative and that can lead to distortions in the data utilized to test abstract theories.

Briefly summarized, the functionalists have tended to begin with general and abstract theories of deviance and then test the theories utilizing a variety of apparently authentic statistics. The critics maintain that this deductive approach glosses over phenomena, including the official rates of occur-

---

[33]See, for example, the Parsonian analysis of male delinquency in Albert K. Cohen, *Delinquent Boys* (New York: Free Press, 1955), especially pp. 156-58.

ence that are subject to a number of biases. They counter the functional method by advocating an inductive approach, which begins with a detailed examination of the social meanings of specific types of deviance.

Conclusion

Durkheim's legacy has involved a view that deviance is normal and positively functional for a society. His followers also view deviance as related, in form, to the social organizational context in which it occurs. Prostitution, witch-trials, political machines and suicides have all been examined from this perspective and their persistence and change have been clarified within this framework. That is, the perspective has provided general insights into the nature of deviant behavior and has offered at least some empirical data that are congruent with the theoretical expectations. This makes it difficult to dismiss the functionalist perspective. Yet, at each step, there is serious criticism of the perspective that is also difficult to dismiss. Its generality, its insensitivity to matters of politics and power, and its naive utilization of official statistics certainly cast some doubt on this theoretical viewpoint.

Given the ferocity of the debate, it is easy to assume that it must finally become an either/or situation: either functionalism or an emphasis upon detailed processes, labeling, and so on. In fact, much of the criticism can be viewed as occurring within, rather than outside of, a functionalist context. In reviewing his own and Gouldner's criticism of Parsons, for example, Rex notes the political naivety and the overemphasis upon stability. What this criticism does, he continues, is expand the possibilities of what can occur "in a functionally articulated system."[34] In other words, it increases the range and versatility of a functional approach, rather than deny the inherent usefulness of such an approach.

This same point can be grasped by further reflection upon the implications of much of the criticism of functionalism. For example, the skeptical view of the use of official statistics by functionalists suggests that the process by which officials make decisions about deviance may be more relevant than the official statistics that are generated. While this charge cuts at a cornerstone of functionalist methodology, it also suggests a perspective that is highly congruent with some basic aspects of functionalism; namely, the view that deviance is functional for a society. In this regard, we have already examined Erikson's position that changing community boundaries trigger "crime waves" a view that emphasizes the arbitrariness of official classifications, and also provides a perspective from which to analyze them historically.

---

[34]John Rex, "The Challenge of Alvin Gouldner," (Review article), *Sociology,* 8 (1974).

The major issue, then, is to try to assess the relative strengths and weaknesses of the functional approach. This is the question we have recurrently encountered in this text. At what point, we continue to ask, does the functional approach move from a net gain to a net deficit? And perhaps most importantly, how can we tell when to apply a functional interpretation? These are the issues we will address in the following chapter.

6

assessing functional interpretations

Throughout the preceding chapters of this book we have repeatedly confronted the question of whether or not a functional interpretation "fits" a specific set of data. Some previously discussed illustrations may facilitate recall. Examples include: the apparent relationship between the specificity of women's sex roles and the prevalence of prostitution in a society; the observation that the educational requirements for many jobs seemed to increase between 1940 and 1970; and so on. In such situations there was ambiguity over whether to offer a functional interpretation. It has been a recurrent question because of conceptual uncertainties regarding how the appropriateness of a functional perspective ought to be assessed. However, this issue far transcends functionalism. It raises the very central question of how the goodness of fit of *any* theoretical perspective is ever shown, in *any* scientific field. Because it presents such a crucial and overriding problem, this general issue will provide the first topic to be addressed in this chapter.

Another major, but more specific, issue involves the relationship between functionality and causality. Contemporary scholars are divided over whether function implies a special type of cause, or whether function and cause are antithetical to each other. These and other alternatives bear directly upon the first issue because there are a number of systematic procedures conventionally employed to assess causality, and it is important to know whether to utilize them or ignore them in assessing functional interpretations. Therefore, the relationship between function and cause will be the second major issue to be discussed. This chapter concludes with a discussion of several procedures that may explicitly be utilized to assess functional interpretations.

Isomorphism's Dual Consequences

Perhaps the most basic task of sociologists is to make sense out of the myriad of mundane events that constitute social life. To make sense out of concrete events (to interpret them) typically requires linking them with abstract principles. If people over age sixty-five are found to be particularly apt to vote for a candidate who speaks in favor of raising social security benefits, for example, then the preferences of the elderly might be explained by the general assertion that people vote in accordance with their perceived self-interests.

This linking of the concrete (elderly voting) with the abstract (pursuit of self-interest) tends to occur either inductively or deductively. In a deductive approach, a previously developed generalization is applied to a new observation. An inductive approach, by contrast, develops generalizations from the specifics. Thus, examination of trade union members, of the affluent, or of minorities might disclose that each group's preferences tend to correspond with its perceptions of the positions of candidates on issues that are particularly salient to that group.

Induction and deduction, in the larger scheme of things, are sequential processes. At one moment, examination of specifics may lead to generalizations. At another time, the generalizations may be applied deductively to specifics that were not previously studied. It is also apparent, however, that an abstract conception may have implicitly guided the first inductive step in the chain just described. Induction and deduction can, therefore, be regarded like the chicken and the egg because of their sequential inter-connectedness.

The fundamental question that confronts both inductive and deductive reasoning involves the goodness of fit, or degree of isomorphism, between the concrete events and the theoretical generalizations. However, it is usually difficult to conclude that there is or is not sufficient isomorphism because specific situations present (or are constructed to present) less than all of the potentially relevant information. The data, as a result, are amenable to a variety of explanations, even if they were implicitly constructed to be "congenial" to a particular theoretical perspective. Therefore, functional interpretations of relationships or trends generate controversy because the specifics to be explained also seem amenable to conflicting theoretical interpretations. Regardless of whether functional interpretations are arrived at inductively or deductively, the basic issue is usually a case of degree of isomorphism.

The consequences of such goodness-of-fit decisions, however, reach beyond how satisfactorily a concrete situation has been explained. These decisions, cumulatively, entail an evaluation of the theory as well. Thus, when an abstract perspective is applied to a specific event there are simultaneously two agenda being pursued, implicitly or explicitly: to explain the specific, and to support the theory by demonstrating its capacity to explain such specifics. In other words, attempts to explain are also exercises in theory testing in which the confirmation of a theory is on the line.[1]

Note that isomorphism is regarded as providing support for a theory rather than proving its "truth." Words such as truth are normally avoided because they connote the possibility of a simple or direct correspondence between theory and reality. Such connotations disregard the complexities that are involved: the subjectivity in evaluating degree of isomorphism, the distortions of data that can arise in their social construction, and so on. In addition, scientists tend to work explicitly with simplified models of reality rather than with the totality of what is "actually real."[2] The tendency, therefore, is to emphasize a theory's explanatory usefulness on the one hand, and its degree of empirical support on the other. Typically, however, both hands are simultaneously involved because of the dual consequences of isomorphic decisions.

Given the profound implications of this decision-making process, there have been surprisingly few efforts to develop systematic procedures that could serve as guides. With respect to functionalist models, the prior efforts can roughly be classified as belonging to one of two types: those paradigms that clarify conceptual issues, but offer no systematic procedures; and those that offer procedures, but tend to overly equate function with cause in too uncritical a manner. Examples of each type are discussed in the following section.

Functional Paradigms

The first type of purported paradigm, best illustrated in Merton's essays, actually presents a clarification of conceptual issues related to Durkheimian (or societal) functionalism rather than a set of procedures for judging

[1] See the seminal essays on the bearing of research to theory, and of theory to research, in Robert K. Merton, *Social Theory and Social Structure* (New York: Free Press, 1957).

[2] The social and natural sciences may differ somewhat in this regard. Theory in social science is meant to apply to reality, Nagel states, but only to limited models of reality in the natural sciences. Ernest Nagel, *The Structure of Science* (New York: Harcourt Brace, 1961). See also, Hubert M. Blalock, Jr., "Some Important Methodological Problems for Sociology," *Sociology and Social Research,* 47 (1963).

the goodness of fit of functional interpretations.[3] Particularly at the time of its initial publication in 1949, Merton's paradigm provided an important critique of conceptual issues. He reminded functionalists not to confuse the objective consequences of behavior with actors' subjective motives; to focus upon patterns; and to be clear about who or what was benefitting from a practice. While Merton's paradigm remains timely today because its dictums continue to be breached in much functionalist writing, it clearly fails to provide a decision-making model. It explicates the questions that ought to be raised in an analysis, but the paradigm offers very little help in deciding whether, in any concrete situation, a functional interpretation is warranted. Merton's model has also been criticized, most conspicuously by Nagel, for failing to provide a sufficiently clear view of the "states" of a social system, and of the causal variables that are relevant to it.[4] This presumed shortcoming is the strong point of the second type of paradigm, and we shall return to Nagel's criticism in our consideration of this approach.

The second type of paradigm has been most extensively developed by biological scientists and is best illustrated in sociology in the writings of Stinchcombe.[5] Coming from a similar causal position, it overcomes Nagel's objections to Merton's paradigm; but it leaves open and problematic the very issues emphasized by Merton. Specifically, this second type of paradigm emphasizes "equifinality," which in biological models refers to the degree to which some final state of a system is a consequence of earlier states.[6] The later development of a fetus, for example, is viewed as a consequence of its earlier development as an embryo. In Stinchcombe's modification, equifinality is taken to mean that the same goal is sought in different ways and/or despite numerous obstacles. The typical social system goal, he assumes, is system maintenance or homeostasis. When this end state is interfered with, or there are deviations from a goal being sought, then adaptive activities are triggered. Assume, for example, that during the first three months of gestation, a human embryo develops through states A and B; then over the next six months the fetus builds onto A and B; developing through stages $C, D,$ and E. Further assume that each stage (as designated by the alphabetical letters) requires the one which precedes it. Thus, if A does not occur, B will not occur; without B there will be no $C, D,$ or $E,$ and so on. Equifinality would be indicated if the fetus were able to overcome obstacles in moving between stages, or if the organism "behaved as though" such efforts were being made. This could be illustrated by the capacity of the fetus to move from C to D

[3]Merton, "Manifest and Latent Functions," in *Social Theory and Social Structure*. See also, William Goode, *Explanations in Social Theory* (New York: Oxford, 1973).

[4]Ernest Nagel, *Logic Without Metaphysics* (New York: Free Press, 1956).

[5]Arthur L. Stinchcombe, "Functional Causal Imagery," in *Constructing Social Theories* (New York: Harcourt Brace, 1968).

[6]See, for example, Ludwig von Bertalanffy, "General Systems Theory," in *System, Change, and Conflict,* eds. N. J. Demerath, III, and Richard A. Peterson (New York: Free Press, 1967).

despite receiving an unusually low nutritional intake; despite not attaining full state C development; etc. This very same line of reasoning can be applied to the socialization process, to large organizations, or other sociological problems. Stinchcombe advocates its application, with equifinality referring to a system's (or subsystem's) efforts to persist and/or attain certain end states.

Stinchcombe's model does offer a basis for deciding whether a functional interpretation is appropriate. Essentially, if the "data suggest equifinality," then functionalist causal imagery is regarded as warranted. This model also meets Nagel's objection by specifying causal relations. Specifically, functional relationships are viewed as special types of causal relationships in which the consequences of behavior (or institutions) are considered their causes in a "circular causal chain." Business organizations, for example, must reduce, or at least cope with, uncertainties in the market place if they are to survive. Their patterns of testing potential markets and ordering inventories for example, can be viewed as ways of reducing uncertainties. Their repeated efforts to attain this goal (i.e., the minimization of uncertainty) is an indication of equifinality. A functional interpretation is warranted, Stinchcombe writes, because the goal being sought determines prior courses of action.

The model's deficiency is its insufficient concern with social versus individual units and mechanisms in its equation of function and cause. Dore illustrates this problem with the study of the interior of a watch.[7] The balance spring, he notes, controls the movement of the balance wheel. This control, it might be inferred, is the function of the spring; but is that its cause? The spring is there because the watchmaker knows its function. Thus, to infer cause from function can require that a knowledgeable agent be identified, and this runs counter both to Durkheimian functional analysis and an emphasis upon the latent functions of institutions or events.[8]

Confusion regarding the equivalence or separateness of function and cause can be traced back at least to Durkheim. As discussed in detail in Chapter 2, he stated that the causes of a phenomenon were independent of its functions, and that, correspondingly, causes and functions must be explained separately. However, Durkheim also viewed functional consequences as causes, arguing that the cause "needs" its effect. Like Stinchcombe, he viewed this need as providing the bond between cause and effect, thereby equating cause and function, despite also insisting on their conceptual separation.

In sum, the conceptual model illustrated in Merton's essays focuses upon questions regarding individual motives and their relationship to social functions. A distinction between the two is strongly emphasized in the Durkheimian tradition, and more recently, by sociologists' focus upon latent

---

[7]Ronald P. Dore, "Function and Cause," *American Sociological Review,* 26 (1961).

[8]Dore concludes that function and cause are not only separate, but that knowing function does not necessarily even clarify cause. On the difference between function and cause, see also, Robert Brown, *Explanation in Social Science* (Chicago: Aldine, 1963).

functions. Within this tradition, however, the relationship between function and cause is ambiguous. By contrast, the equifinal models regard functionality as a specific type of causality. John Stuart Mill's old dictum that causes temporally precede effects is simply conceptually reversed in the equifinal notion of circular causality. However, "purposefulness" is rather indiscriminantly introduced by these models. If the motivated behavior of individuals must be assumed in order to conceptualize functions in causal terms, then this notion of causality is incongruent with the Durkheimian tradition. It is not surprising from this point of view that many of Stinchcombe's illustrations are drawn from Malinowski's writings, which emphasized an individualistic rather than societal form of functionalism.

On the other hand, there is nothing inconsistent with assuming purposefulness on the part of the system, with maintenance of homeostasis being inferred as the ultimate goal. Thus, it is society that is regarded as possessing the "unseen hand," and the subjective motives of participants are not expected to concur. Stinchcome, in fact, makes a similar point, arguing that inconsistency among the stated motives of participants is one indication of an equifinal structure. A major difficulty with this conception, however, is that society is prone to be viewed from an individualistic metaphor; that is, needs, motives, and the like are often attributed to society, and this may well involve reification. In general, reification entails assigning substance to an abstraction; thus, the problem hinges upon whether or not society is to be regarded as a "thing." Even if it is, however, it is not clear whether it is appropriate to attribute individual qualities, such as motives, to it. This may entail a kind of reductionism in reverse.

The compatibilty of function and cause is an important issue because both the logic of and procedures for determining causality have been developed over the years. If functional relationships could be regarded as a special type of causal relationship, then causal procedures could be utilized in assessing the appropriateness of functional interpretations. There are, however, many types of functional theories and we should not anticipate that they will all be equally compatible with the assumptions relating to causality. Prior to examining these variations, it will be useful to briefly consider some of the assumptions and procedures that have been conventionally employed in causal approaches.

Causal and Functional Models

The very notion of cause (like the notion of function) is highly problematic. It has no widely shared, unambiguous meaning, and causality is virtually impossible to demonstrate unequivocably. However, most causal models do share certain assumptions, and there are several other issues that all of them address even when they do not make the same assumptions. In the following pages we will discuss four of these core issues: patterns, spuriousness, symmetry, and externality. Each of these issues will be examined in relation to both causal and functional perspectives.

patterns

Underlying any scientific conception of causality is the assumption that reality contains relatively invariant regularities, or that models of reality can be constructed from which such constant patterns can be inferred. It is more specifically assumed that these patterns involve causes and effects that are uniformly tied to each other. A stone hitting the water, for example, is the cause of the ensuing splash; and the impact and the splash are a recurrent causal sequence.

The assumption of causal patterns are a refutation of David Hume's philosophy that nothing exists, either in nature or our experience, except one object following another.[9] Social scientists, among others, have responded to this view by arguing that our entire experience of the world is molded by causal notions and that this total experience would be meaningless without it. For example, we cannot conceive of uncaused change, MacIver has stated, because it would signify that time was random and meaningless. It would be like, "the moving picture on a screen that can be run backward as well as forward, which can be slowed or accelerated"; however, we perceive continuity in change because "the concept of causation reigns over our experience" and converts change into a patterned "one-way road from the past to the future."[10]

The possibility of isolating and describing patterns, either because they exist in reality or because they can be inferred from constructed models, is an assumption that is generally shared in functional as well as causal analysis. It was the first step, for example, in Merton's functional paradigm. Whether the patterns must be the result of causes and effects is another matter, however.

spuriousness

Strictly speaking it is not possible to prove that one variable causes another. All the procedures and techniques of any science are geared to prove that a causal relationship does not exist. When repeated efforts to prove noncausality fail, the failures are interpreted as the confirmation of a causal relationship. When the attempt to disprove causality is successful, the relationship is termed spurious. Thus, imagine a dichotomy of all relationships that could be empirically observed. One category would contain all the rela-

---

[9]David Hume, "Of the Idea of Necessary Connection," in *The Scottish Moralists*, ed. Louis Schneider (Chicago: University of Chicago Press, 1967).

[10]R. M. MacIver, *Social Causation* (New York: Harper, 1964), pp. 6-7.

tionships in which a causal tie between the variables could not be rejected, and it would be termed causal. All those relationships for which causality is rejected would then be placed into a spurious category. Finally, it is also possible to refer to spuriousness, involving the degree to which a relationship is non-causal.

Intuition, logic, or insight typically provides a preliminary basis for differentiating between spurious and potentially causal relationships. For example, in northern provinces of Russia, Sorokin reports that peasants wear tall, warm boots, and that they gather in the evening to present plays and sing. If the provinces of Russia were taken as cases, it would presumably be possible to observe an association between warm boots and evening festivities: both would be present together in the provinces of the North and both would be absent in the South. They vary together, but their association is probably spurious, Sorokin concludes, because there is no meaningful or logical tie between the variables. Their association could be viewed as mere happenstance, or as due to the common association of each with another factor, namely, long and cold winters.[11]

There are a number of formal techniques by which spuriousness can be detected.[12] They typically hold constant the factor or factors common to both variables, and indicate the degree of spuriousness by noting the degree to which the originally observed relationship then declines. In the above example, provinces might be classified according to the severity of their winters, and the original association re-examined within each category. Controlling for the severity of winter would probably eliminate all connections between boot length and evening festivities, indicating that it was entirely spurious. In other words, there is no direct association between these variables. The "apparent" association is due totally to their association with the third variable, the location of provinces.

In most actual studies, unlike the hypothetical study of Russian provinces, detecting spuriousness is very problematic. There are usually a limited number of measured variables, but a never-ending list of potentially spurious variables can be conceptualized. At some point a theorist must arbitrarily stop because moving from an observed relationship to inferred causality is mired in a potentially never-ending search.[13]

Despite these practical limitations, eliminating spurious relationships is at least consistently emphasized as an ideal goal in causal models. Spuriousness may present a more problematic issue in functional analyses, though. If all components (i.e., variables) in a system are assumed to be highly interre-

[11]Pitrim A. Sorokin, *Social and Cultural Dynamics* (New York: Bedminister Press, 1937).

[12]See Hubert M. Blalock, Jr., *Causal Inferences in Nonexperimental Research* (Chapel Hill: University of North Carolina, 1964).

[13]See the discussion of "infinite regress," in George V. Zito, *Methodology and Meanings* (New York: Praeger, 1975).

lated, then all variables serve as the third variable in the relationships among all other variables. Methodologically it would not be possible, in such a situation, to remove all the shared variation that is spurious.[14] Further, to attempt to remove the shared variation that is spurious—from a causal perspective— might distort the basic features of such a functionally viewed system.[15]

symmetry

A simple relationship means only that variables vary together. Even if the relationship is apparently not spurious, many causal models do not view it in causal terms unless the direction of the relationship can be inferred. Directionality can be conceptualized as the asymmetry of a relationship, and it is illustrated by a situation in which variable X produces variable Y, but variable Y does not produce variable X. For example, poor weather leads to small wheat crops, but not vice versa.[16] If Y also produces X, then the relationship is regarded as symmetrical, and therefore is sometimes regarded as non-causal. Historically, a major conceptual difference between function and cause is that functional interrelatedness typically implies symmetry while causality sometimes implies asymmetry.

Causal models are often predicated upon asymmetrical assumptions for methodological convenience; identifying effects involves special problems if a causal relationship is viewed as going in both directions. And conceptually, the notion of X producing Y, but not the reverse, is isomorphic to some widely shared interpretations of the meaning of causality.

For both of the above reasons, there have been many recent efforts in sociology to work with asymmetrical causal models. Special recursive systems are an excellent example.[17] They involve placing variables into a set of recursive equations that, by definition, allow for no "feed-back"; that is, variables cannot be both cause and effect of each other. This type of causal model, it should be noted, rules out the kinds of reciprocal efforts that have historically been associated with functionalism.[18]

[14]The problem of under-identifying the model would be insurmountable.

[15]See L. Keith Miller, "A Methodological Note," in Alvin W. Gouldner and Richard A. Peterson, *Technology and the Moral Order* (Indianapolis: Bobbs-Merrill, 1962).

[16]Herbert A. Simon, "Causal Ordering and Identifiability," in *Essays on the Structure of Social Science Models,* eds. Albert Ando, et al. (Cambridge: M.I.T. Press, 1963), Chap. 2.

[17]Hubert M. Blalock, Jr., "Theory Building and Causal Inferences," in Hubert M. Blalock and Ann B. Blalock, *Methodology in Social Research* (New York: McGraw-Hill, 1968), Chap. 5.

[18]A favored recent technique, involving most of these assumptions, is path analysis. For further discussion of this technique, see Otis D. Duncan, "Path Analysis: Sociological Examples," *American Journal of Sociology,* 72 (1966).

externality

A final assumption common to many causal models, and particularly to those developed in econometrics, is the possibility of theoretically differentiating between endogenous and exogenous variables. The latter are, at least theoretically, located outside of the system in which the other variables are located. Thus, the price of wheat, from our previous illustration, would be an endogenous variable; part of the pricing system. The weather, by contrast, would be an exogenous variable; that is, theoretically located outside of the system.[19]

Placing variables into endogenous and exogenous categories, as already suggested, is an arbitrary classification that is derived from a theoretical scheme. Almost any given system or subsystem could be conceptualized more broadly, thereby expanding system boundaries to include formerly exogenous variables. Even though it is arbitrary and fluctuating, however, many causal models include such a categorization because it is easier to estimate the causal effects of designated exogenous variables upon the endogenous variables of a system. Correspondingly, assumptions of asymmetry and recursiveness are facilitated.

From a functionalist perspective, objections to the exogenous classification are similar to those previously discussed in regard to spuriousness and asymmetry. Specifically, many functional theories, such as Parsons's, so broadly define systems that very little is left outside of them to serve as external, or exogenous, variables. It is true that cultural, social, and psychological systems are differentiated; but the interpenetrations among these systems are usually assumed to be too great to permit the identification of exogenous variables. Other theories, such as Durkheim's, are explicit —conceptually, at least—about system boundaries. However, between-system analyses are specifically condemned; for example, social facts are not to be accounted for by psychological facts.

conclusions

Historically, the greatest opposition to merging functional and causal analyses was probably expressed by Pareto. His functional model emphasized the necessity of focusing upon patterns or uniformities; but he rejected causal interpretations as overly simplified because of assumptions like asymmetry and recursiveness (though he did not use these terms). Pareto regarded the form of society as a product of many elements, but societal form was also

[19]Simon, "Causal Ordering and Identifiability."

viewed as acting upon these elements.[20] Thus, he concluded that it was neither possible nor fruitful to identify "conventional" causal relationships.[21]

The historical opposition of functionalists to causal models, it must be emphasized, is really directed at certain assumptions that are not absolute requirements of causal analysis. Recursive models are only one type of causal model; asymmetry need not be assumed; and exogenous variables are not indispensible. However, while causal analyses can be performed without any of these assumptions, their absence does introduce difficult methodological problems. In addition, without these assumptions it can be conceptually unclear what cause means.

Approaching the opposition between functional and causal models from another angle, it must also be emphasized that their presumed incongruity has been stressed in relation to extreme functional perspectives, which rather loosely state that everything is related to everything else. There have also been long-standing efforts to reconcile the two by modifying this extreme functional view. MacIver, for example, has argued that within the context of total (within-system) interdependence, "there are more specific and limited interdependences."[22] They are amenable to causal analysis, he continued, if external, causal variables are identified. In more contemporary functional terms, MacIver was urging that society, as a system, be conceptually differentiated into several subsystems.[23] The externality assumption could then more easily be met as causal relationships are sought between subsystems, or between variables thought to be primarily located in different subsystems. Asymmetry and recursiveness have also become more congruent with functional conceptions when theorists argue that the components of systems (i.e., subsystems) may be regarded as engaged in relationships of unequal reciprocity.[24] This involves viewing subsystem A as influencing subsystem B more than subsystem B influences subsystem A. However, while the notion of unequal reciprocity is somewhat akin to asymmetry in a causal sense, it is doubtful that they can be equated without resulting in conceptual distortion. To what degree, for example, can the notion of unequal reciprocity be translated into a one-directional, asymmetrical relationship in a recursive equation?

Finally, it may be possible to reconcile a functional perspective with the most demanding causal assumptions by focusing upon system "states." (A failure to do so, it will be recalled, was Nagel's criticism of Merton's paradigm.) Theoretically, such states refer to a total system at any given point

[20]Pareto's conception of both societal form and social forces, however, included variables considered by Durkheim to be outside the realm of the social. Vilfredo Pareto, *The Mind and Society* (New York: Dover, 1963), Chap. 2. (Originally published in 1916).

[21]See also the more recent statement by Von Bertalanffy, "General Systems Theory."

[22]MacIver, *Social Causation,* p. 254.

[23]There have been a number of theoretical categorizations of this type; most notably, see Talcott Parsons and Neil J. Smelser, *Economy and Society* (New York: Free Press, 1959).

[24]See, for example, Gouldner and Peterson, *Technology and the Moral Order.*

in time. The system could be regarded functionally as composed of mutually interacting elements whose interrelationships may not be amenable to conventional causal analysis. However, the resultant outcomes of these processes —system states—may be amenable to such an analysis, utilizing, for example, an equifinal model.

Let us now summarize this discussion by returning to the question that prompted it. Our initial concern was with how to assess the appropriateness of functional interpretations. One possible strategy that we investigated was the viewing of a functional model as a type of causal model. Correspondingly, we examined some causal assumptions, questioning their congruence with functional perspectives. While the two were found not to be inherently contradictory, there were some instances when their compatability may have been strained.

I would emphasize two conclusions. First, causal models cover such a wide spectrum that to claim functional explanations are in fact a type of causal explanation helps to clarify little. Second, to accept this overlap would provide little help with respect to knowing when a functional interpretation is appropriate. In other words, only those observed relationships where causality could not be rejected might be considered "eligible" for functional interpretations. However, such interpretations would remain arbitrary; no implications unique to a functional perspective are directly tested.[25] Therefore, our attention now turns directly to the question: what besides, or in addition to, causality should be indicated before a functional interpretation is justified?[26]

Special Mechanisms and Functional Efficacy

All functional theories emphasize the consequences of a relationship or a practice for the maintenance, homeostatis, or persistence of a system (or subsystem). Even if a relationship is causal, a functional interpretation is not warranted unless the relationship has such consequences (i.e., functions or dysfunctions, though the latter are not examined here). Therefore, we may conceptually allow for the possibility of causal relationships that are largely irrelevant to system (or even subsystem) maintenance, and thus, are of little functional concern. To separate function and cause in this manner requires only that the complete functional integration of a society not be assumed, and

[25]This objection is less applicable to an equifinal analysis of system states. Here there are, at least, some "clues," such as the conflicting motives of participants, and the fact that cause and effect relationships follow a relatively unique temporal pattern. See Stinchcombe, "Functional Causal Imagery."

[26]Note that the emphasis here is upon the appropriateness of a functional interpretation and not upon demonstrating that such an interpretation fits best. The latter, as discussed in previous chapters, requires a critical test.

[27]See Sorokin, *Social and Cultural Dynamics*.

that even institutionalized patterns be regarded as capable of possessing a degree of autonomy.[27]

Despite their previously noted differences, the paradigms associated with Merton and Stinchcombe share, at least implicitly, the simple yet important notion that any relationship or pattern that is presumed to be functional must possess some degree of functional efficacy. Unlike functions, per se, which can refer to purposes or motives, the notion of functional efficacy emphasizes actual consequences. This difference is illustrated in the following two statements. The first, in typical functionalist form, states: the function of adrenalin is to counteract the effects of trauma. Thus, adrenalin is explained by its presumed contribution to a system. To report that adrenalin counteracts trauma, however, is a quite different statement.[28] The latter, because of its emphasis upon actual consequences, reduces the ambiguity of the term "function" by removing from consideration such notions as motives or intentions. In the following pages we shall describe a paradigm that focuses upon relationships rather than individual components (such as adrenalin) and that examines the consequences of these relationships. Our basic contention will be that functional interpretations are not warranted unless the relationships can be shown to contribute to system maintenance, facilitate goal attainment, or reduce strains or tensions.

Associated with the notion of functional efficacy is an emphasis upon specific ends that are "threatened" by concrete circumstances. In his previously discussed analysis of political machines, for example, Merton viewed voter apathy as disrupting normally functioning political institutions. When such problematic conditions occur, the functionalist is regarded by Merton as "sensitized" to look for the possible effects of "compensating mechanisms." Similarly, in Stinchcombe's view, equifinality is indicated when certain ends are repeatedly sought, despite the obstacles that may be delaying their attainment.

This focus upon specific ends, concrete obstacles, and the functional efficacy of compensating mechanisms stands in marked contrast to an earlier theoretical emphasis upon functional prerequisites. In this earlier approach, various needs were viewed as built into any system, as being constant features. The notion of functional efficacy was correspondingly unemphasized because the sheer persistence of any system was regarded as de facto evidence that institutions and practices must be functioning to meet these invariant system needs.[29]

[28]This illustration is taken from Brown, *Explanation in Social Science*. However, using different terminology, Brown proposes that it illustrates the difference between function and cause.

[29]For further discussion of the limitations of the functional prerequisite approach, see Goode, *Explanations in Social Theory;* and Mark Abrahamson, "On the Structural-Functional Theory of Development," in *Social Development,* ed. Manfred Stanley (New York: Basic Books, 1972).

To be more specific, we can outline the following three steps as a paradigm for assessing the adequacy of functional interpretations of observed relationships:

1) Identify the problematic conditions that make special mechanisms efficacious for the attainment of certain ends.
2) Demonstrate that the ends are more likely to be attained, despite problematic conditions, when the special mechanisms are present.
3) Show that attainment of the ends is not enhanced by the special mechanisms when the problematic conditions are absent.

In order to illustrate the usefulness of these steps in indicating when functional interpretations are warranted, they will be applied to several previously observed and functionally interpreted relationships. Some unavailable data preclude the application of all the steps in every example to be considered. However, our primary concern lies more with illustrating the heuristic value of the paradigm than with providing a definitive assessment of the specific relationships.

initiation ceremonies

In their previously discussed study of male initiation ceremonies, Whiting and several of his colleagues proposed to explain, "the function of male initiation rites which accounts for the presence of these rites in some societies and the absence of them in others."[30] It should be recalled that in their cross-cultural survey they reported that the ceremonies were most likely to be present in societies that also possessed certain infant socialization practices: an exclusive mother-son sleeping arrangement and/or a post partum sex taboo. Unless compensated for, these practices were assumed to result in males' psychological over-dependence upon females and their subsequent development of inappropriate sex identifications. Adult male performance of various sexual and sex-role-related activities was expected to be impaired by anxieties unless the society held initiation ceremonies to redirect male identifications.

The theoretical underpinning of this explanation is highly psychoanalytic; but the basic argument can readily be translated into the terms of our functional paradigm. Specifically, the early socialization practices may be identified as the problematic condition; initiation ceremonies may be regarded as the special mechanism; and male anxiety reduction as the end to be attained.

The first step is to ascertain whether the early socialization practices do, in fact, produce male anxieties. A causal relationship between them was

[30]John W. M. Whiting, et al., "The Function of Male Initiation Ceremonies at Puberty," *Readings in Social Psychology,* eds. Eleanor Macoby, et. al. (New York: Holt, Rinehart, 1958), p. 360.

assumed by the investigators. It was not tested, however, as their study included no measure of anxiety. To make assessment possible, I constructed a cross-cultural male anxiety scale from data that had been reported in a number of other studies. The scale contains four items purporting to measure: sexual disability;[31] castration anxiety,[32] bellicosity;[33] and emphasis upon military glory.[34] The first two items focus directly upon sexual anxieties, while the last two items appear to involve non-sexual ways of displaying masculinity. The response patterns of the four items are congruent with each other, however, indicating that they can be combined into a single composite measure.[35] The preliterate societies that were given scores in all the prior studies were utilized as the sample.

When mean scores on the male anxiety scale are examined in societies with and without the early socialization practices, the hypothesized relationship is observed to be very strong. This association, it will be recalled, is expected to occur only in societies without male initiation ceremonies; that is, without the presumably compensating mechanism. Therefore, the following figures apply only to the twenty-three sampled societies that lack initiation ceremonies. Specifically, the mean male anxiety scale score in societies that possess either of the early socialization practices is 34.7. By contrast, in the societies without either practice, the anxiety scale mean is only 14.2.

The strong and significant ($F = 6.72, p < .05$) association between the socialization practices and male anxiety increases confidence that the specially contructed anxiety scale is an indicator of the kind of anxieties theoretically implied by Whiting and his colleagues. However, the association, per se, does not provide sufficient evidence that the variables are causally connected. As previously discussed, a causal inference requires additional support, which will not be sought here.

Step two of the functional paradigm requires the demonstration that some ends are more likely to be attained, despite the problematic condition, when the special mechanism is present. In this example, the specific hypothesis states that among societies having either socialization practice, anxiety scores will be lower in those that have initiation ceremonies than those that do not. The difference in scale means (31.6 and 34.7, respectively) is in the expected direction, but it is very small and not statistically significant. Especially when compared to the large difference in means associated with the presence or absence of the socialization practices, the amount of anxiety reduction that might be attributed to the initiation ceremonies seems trivial.

[31]John K. Harley, *Adolescent Youths in Peer Groups* (Unpublished Ph.D. dissertation, Harvard University, 1963).

[32]William H. Stephens, *The Oedipus Complex* (New York: Free Press, 1962).

[33]Philip Slater, *Coding Guide for the Cross-Cultural Study of Narcissism* (Unpublished, Brandeis University, 1964).

[34]*Ibid.*

[35]Scalability of the four items is indicated by their (Cornell technique) reproducibility of 0.91. Maximum Error Reproducibility is also an acceptably high, 0.75.

This finding leads to the conclusion that Whiting and his associates offered an inappropriate functional interpretation of the relationship between the socialization practices and initiation ceremonies. It may be that some other functional interpretation would be warranted; of course, it may also be that the relationship should be interpreted according to a non-functional perspective. While further assessment is required to indicate which of these two alternatives should be pursued, step two of the paradigm at least suggests that the functional explanation provided by Whiting and his associates does not fit. This failure to indicate efficacy makes further application of the paradigm unnecessary. Thus, in order to illustrate the third step, we now turn to yet another example.

high gods

In testing hypotheses relating to Durkheim's view of social organization and religious beliefs, Swanson offers a number of functional interpretations.[36] One of the more interesting interpretations involves the relationship between the number of sovereign groups in a society and the presence of a single high God. These sovereign groups possess jurisdiction over activities in a society. The more that are present, the more "fragmented" is the society. Swanson observed that a single high God, as opposed to numerous lower deities, is more likely to occur when there are many sovereign groups. He explained the relationship by suggesting the unifying function of a high God. Unlike lower deities, who are perceived as specialized, people's belief in a single high God is more capable of unifying an entire society. When the society must respond in unison to a threat (i.e., external attack), but its ability to do so might otherwise be impaired by fragmentation, belief in the single high God is presumed to provide a particularly effective basis of integration. Hence, fragmentation is the problematic condition, belief in a high God is the special mechanism, and effective unification is the end state.

To test the assertion that fragmentation is the problematic condition would require demonstration of a negative causal relationship between fragmentation and social integration. This first step was illustrated in the preceding analysis of male anxiety by showing that such anxiety was apparently produced by the two socialization practices. Therefore, let us assume that the necessary causal relationship could be demonstrated and proceed to the following steps. The next step hypothesizes that the end state is more likely to be attained when the special mechanism (in this situation, belief in a high God) is present. One possible end suggested by Swanson is war-related success. That is, the military efforts of more unified societies might be more successful, and their potential enemies might feel more intimidated by their potential

[36]Guy Swanson, *The Birth of the Gods* (Ann Arbor: University of Michigan Press, 1960).

prowess. A possible operationalization of this end state is a bellicosity measure developed by Slater[37] and later extended by Textor.[38]

The specific hypothesis of step two may now be stated as follows: the bellicosity scores of fragmented societies will be higher among those in which there is belief in a high God. Looking at all fifty-five preliterate societies for which data is available provides clear support for the hypothesis. In the thirty-one fragmented societies characterized by belief in a single high God, nearly two-thirds of the societies score high in bellicosity. By contrast, in the twenty-four fragmented societies that do not have a belief in a single high God, only one-third are high in bellicosity ($X^2 = 4.33, p < .05$).

This finding supports the functional efficacy assertion; however, the efficacy of belief in a high God must be further tied to the problematic condition (i.e., fragmentation), and this introduces step three. Specifically, it is hypothesized that there is no association between bellicosity and a high God in non-fragmented societies. Examination of the seventeen non-fragmented societies in the sample shows that, in fact, there is no association. High bellicosity occurs in about 40 percent of these societies, whether there is a belief in a high God or not.

Summary and Conclusion

This chapter has posed a question concerning how to decide whether or not a functional interpretation "fits" an observed relationship. A three-step paradigm, focusing upon functional efficacy in relation to problematic conditions, was presented as a procedure for answering this question. The minimal requirements of any such testing procedure is that it be capable of differentiating. If its application indicated that functional interpretations were always or never warranted, then it would contribute nothing to the decision-making process. From the examples presented, however, it does appear capable, at least, of differentiating.

The need for such a paradigm seems apparent. Prior studies have tended to focus solely upon two variables. Examples of such studies are those that focused on early socialization practices and initiation ceremonies. If the hypothesized relationship was obtained, a functional interpretation was offered. There were often theoretical precedents for such interpretations, such as Durkheim for Swanson. However, we have argued that an observed relationship between two variables is insufficient evidence, and have presented the three-step paradigm to suggest what additional evidence might be necessary.

[37] Slater, *Coding Guide*.
[38] Robert B. Textor, *A Cross-Cultural Survey* (New Haven: HRAF Press, 1967).

One serious problem associated with the prior studies' sole focus upon two variables is that they implied absurd causal connections. Thus, early socialization practices should not have been conceptualized as causing initiation ceremonies; fragmentation should not have been conceptualized as causing the belief in a high God. Causality was apparently implied, though, because such studies evoked responses which argued that the reported relationships were spurious.[39] However, the studies typically neglected the causal tie in the larger functional framework and the spuriousness issue was then misdirected.

With respect to causality, my argument is that functional interpretations often begin with a causal inference that ties a problematic condition to a problematic state, such as high anxiety or low integration. Neither the problematic condition nor its resultant state should necessarily be viewed as causing the special compensatory mechanism. That mechanism may be the result of a long causal chain in which there is only a weak direct tie between it and the problematic condition; or, it may simply develop in response to quite different conditions. The question of whether it is functional, given the problematic condition, should not be confused with questions of causality.

By widespread application of this three-step paradigm, it would be possible to discuss, from an empirical basis, how typical it was for social practices or institutions to be functionally efficacious. If efficacy was typical, then the more abstract conceptions of functionalism—integration, homeostasis, and the like—would seem more warranted. It is also possible, perhaps highly probable, that inferences of functional efficacy would receive support in some realms or contexts, but not in others. That would also be a welcome finding because it too would result in a way of limiting the arbitrary application of functional interpretations.[40]

Finally, I recognize the possibility that the paradigm presented here might not be strong enough to face the task it was meant to pursue, or that some other paradigm might be capable of performing better. I am convinced, however, that some set of assessment procedures is necessary. Otherwise, large numbers of sociologists will continue to extensively apply functional interpretations in an indiscriminate manner, and many others will continue to reject all functional interpretations out of hand. These differences have generated ideological controversies that have led to reflective insights into the nature of the sociological enterprise. Despite such insights, however, ideological grounds must not provide the criteria by which the conflict is resolved.

---

[39]With respect to early socialization practices and male initiation ceremonies, for example, see the argument by Y. A. Cohen, "The Establishment of Identity in a Social Nexus," *American Anthropologist,* 66 (1964).

[40]It should also be recalled that the three-step procedure presents a way of testing only the appropriateness of a functional interpretation; it does not address the question of whether functional interpretations fit better than others.

index

111